Hunter and Hunted

Relationships between carnivores and people

HANS KRUUK
*Centre for Ecology and Hydrology
and University of Aberdeen*

Drawings by
DIANA E. BROWN

CAMBRIDGE
UNIVERSITY PRESS

PUBLISHED BY THE PRESS SYNDICATE OF THE UNIVERSITY OF CAMBRIDGE
The Pitt Building, Trumpington Street, Cambridge, United Kingdom

CAMBRIDGE UNIVERSITY PRESS
The Edinburgh Building, Cambridge CB2 2RU, UK
40 West 20th Street, New York, NY 10011-4211, USA
477 Williamstown Road, Port Melbourne, VIC 3207, Australia
Ruiz de Alarcón 13, 28014 Madrid, Spain
Dock House, The Waterfront, Cape Town 8001, South Africa

http://www.cambridge.org

First published 2002

Printed in the United Kingdom at the University Press, Cambridge

Typefaces Swift 9/13 pt and Zapf Chancery   *System* LaTeX 2_ε [TB]

A catalogue record for this book is available from the British Library

Library of Congress Cataloguing in Publication data
Kruuk, H. (Hans)
Hunter and hunted: relationships between carnivores and people /
Hans Kruuk; drawings by Diana E. Brown.
    p. cm.
Includes bibliographical references (p.).
ISBN 0 521 81410 3 (hb) – ISBN 0 521 89109 4 (pb)
1. Human–animal relationships.   2. Carnivora – Behavior.   I. Title.
QL85 .K78   2002
599.7′15 – dc21   2002016589

ISBN 0 521 81410 3 hardback
ISBN 0 521 89109 4 paperback

To Jane

Contents

Preface

Watching animals in the wild has occupied much of my life so far, and carnivores were usually central to this. I had the excitement of living amongst foxes, hyaenas, lions, badgers and many others, and almost daily I can watch otters catch their fish just a short distance from my house. I want to keep it that way. More and more I am aware of how privileged I am, in being able to spend time with these wonderful animals in their natural haunts. More and more, also, I am aware of the desperate need to preserve a place for them in our world.

Carnivores are often unpopular, because of the damage they may do to livestock, because of a threat to our person, or because they compete with us over game. We also live in this world with an instinctive, anti-carnivore behaviour to protect ourselves. But at the same time we have an extraordinary relationship with the animals as pets, and we exploit them in several ways. Many of us recognize the wonderful beauty of wild carnivores, and their important role in natural ecosystems.

In this book I attempt to analyse these enigmatic and contradictory relationships, and I try to explain our fascination for the dangerous beauties. Hopefully, this will help to improve the chances of their long-term survival, which is what I especially care about. But the analysis of our relationships with attractive predators and competitors may also help us to understand ourselves. It enables us to see mankind as another species, as another mammal, with its own characteristic anti-predator behaviour that has evolved in response to particular ecological threats and requirements. Uniquely, our inherited anti-predator system is augmented and modified by culture, which, in this context, acts within our species as a highly important process of communication of individual experiences with these animals.

My interests in carnivores and anti-predator behaviour owe much to the late Niko Tinbergen at Oxford, who as a wonderful naturalist and teacher of ethology opened my eyes and those of many others. Later, this process was developed further in Africa, guided by my late friends John Owen, Hugh Lamprey and Myles Turner, and I owe a tremendous debt of gratitude to my many students and colleagues who were involved in the various projects. I am grateful to Steve Albon and staff at the Institute of Terrestrial Ecology (now the Centre for Ecology and Hydrology) in Banchory, Scotland, who in many ways enabled the writing of this book, and to Diana Brown for her inspiring drawings. Loeske and Jane Kruuk, Matt Gompper, Joshua Ginsberg and an unknown referee ironed out many of the mistakes and deficiencies in my writings, for which I thank them deeply. Especially, I acknowledge the helpful interest, tolerance and love from my family, Jane, Loeske and Johnny. Alec Birkbeck, Sim Broekhuizen, Ray Hewson, Andreas Krantz and Ilan Rootsi provided references and ideas.

Spotted hyaena, meerkat, lion, black-backed jackal, raccoon, stoat, brown bears

1

Turning the other cheek

CONUNDRUM

This is a story about competition, about predation, and about fear and attraction and beauty. It is the tale of our relationship with carnivores, both wild and tame, as hunters and pets, killers and scavengers. The book is about our own behaviour as well as about theirs. I will be discussing the ramifications of a simple question: why do we like carnivores so much, and why are we so totally fascinated by animals that are designed to be our enemies?

An early morning breaks over the huge, open grassland plain of the Serengeti, in East Africa. I am driving well away from any road, and my world is a vast expanse even beyond the horizon. The earth is just beginning to breathe in the sunlight, and small birds are stirring. Black dots appear ahead of me, some turning into ostriches, some into wildebeest. I stop, and I listen to the soft and distant grunts from the herd. It is a scene of total peace and expectation.

Beyond the wildebeest something stirs. Gazelles are running, and the wildebeest stop grunting. Lithe sinuous forms appear from the distance, a pack of sixteen African wild dogs, silent and fast. They create chaos all around them, and the wildebeest wheel and flee, bunching up with whisking tails. The dogs are criss-crossing in the turmoil.

One wildebeest cow separates from the herd, with her a calf next to her like a small motorcycle sidecar, both going as fast as their legs can carry them. One dog is behind them, then several. I drive alongside some 20 m away, but none of the animals take any notice of me and my Land-Rover. The first dog nips at the flank of the calf, then at the mother. The cow wheels, and there is menace all around her. Dogs bite the calf, the mother attacks, she is bitten but she can defend herself. The calf has no chance. Later, the lone wildebeest cow stands at a distance, watching the steaming heap of ravenous dogs tearing at the small body that so recently ran beside her.

It was just one, single incident. There were thousands of wildebeest, gazelles and other animals, they were grazing, socializing, defending territories and playing the mating game. But in all this it was the predator and its kill that drew every ounce of my attention. It was a compulsion stronger than myself, and I had to admit to a distinct quiver of excitement whilst making my notes.

I have spent most of my life studying animals, and especially carnivores. I studied predation by foxes and stoats on gulls in Britain, I lived for many years in Africa where I spent most of my time watching hyaenas, lions, wild dogs and many other predators. Over many years I watched badgers at night and otters in the daytime, and it is no overstatement when I describe myself as a carnivore addict.

I may be involved with these animals more than an average person, but some of the same addiction throbs in the veins of many of us. Visitors to African national parks want lions, leopards and cheetahs, and when you see a huddle of cars in the Serengeti there will be a big cat in the centre. Take children to a zoo and they will make a beeline for the tigers, lions and wolves. Many a natural history programme on TV will have a predator in its climax. Our fairy tales and coats of arms bristle with carnivore violence, and pandas and tigers head the conservation urge.

Many of these animals are lethal. They kill many people in developing countries and they would kill people in developed countries if they had a chance. They murder our livestock, they take our game and they give us diseases. Yet those of us from developed countries think carnivores are wonderful, magnificent and almost unbelievably attractive, and we spend millions on their conservation. Even in the developing world many people are fascinated, and maybe even proud of them.

There is an inherent contradiction in this, which I want to explore. The questions of the why and how of our relationship with carnivores are valid ones, because they seek to understand our own instinctive fears and our nightmares, and our preoccupation with the issue of violence. At the same time, the answers are relevant to some of the species that face imminent extinction.

This is the *raison d'être* for this book, which is different from others that have described carnivores and their behaviour or ecology, or the

damage done to us by predators. I want to look here at the relationships between them and us in the same way as I would study the
predator–prey relationships between wild animals. I want to see the
ecological aspects, the actual and potential influence of carnivores over
humankind and of humankind over carnivores, including predation as
well as competition and beneficial effects. Against that background we
need to evaluate behaviour – our own anti-predator reactions to these
animals – in order to study how effective this behaviour is, and what
it does to the animals.

In the following pages I want to approach this from several different angles. I will start from a vantage point, surveying the multitude of
carnivore species, and I will describe some of the order and uniformity
in the variety. Such order is not confined to appearances: it is also there
in their behaviour, in social life and hunting. This point is important
for the perception of carnivores by our own species, because we tend
to lump like with like. Similarities, whether real or perceived, are the
basis for our prejudices.

I will then move on to what affects us directly, to the mechanism
of the relationship and aspects of carnivore behaviour that are involved
in causing damage to the human race. Several carnivores are maneaters,
and I will present the case against them in detail. Many of them also
cause substantial damage to our livestock and to the game we covet, and
substantial financial resources have to be used against them, adding to
a long list of other charges. It is not difficult to demonstrate that this
damaging relationship between the animals and our own species goes
back right to the beginning of our very first steps on this planet.

However, carnivores also have another side. Their story is a litany
of contrasts, because in our present-day society we need them. We
derive many benefits from pets and working animals, we even refer
to 'man's best friend'. There is a worldwide trade in furs, and carnivores provide medicine, food and 'sport'. Also, their mere presence can
be seen as a benefit to us: we find them beautiful, exciting, the epitome
of everything that is wild.

Against such a background of debits and credits, I describe in
some detail in the following chapters how we, as a species, react to the
animals. At its most basic level, human behaviour towards carnivores
often contains clear elements of fear and of aggression, and of strong
curiosity. In this, people are not alone, for these same elements come
back in birds, in gulls mobbing a fox, and in the many other mammals
that share their living space with predators that have designs on their
lives or those of their offspring. It is sometimes easier to get an objective

insight into the behaviour of wild animals than to rationalize our own reactions, so I will describe the anti-predator behaviour of birds and others, to arrive at an understanding which can then make a small contribution to our knowledge of ourselves. Our own anti-predator behaviour has much in common with that of others – of wild birds and mammals.

However, there is more in our reactions to carnivores than just basic, instinctive anti-predator behaviour. We experience appreciation of a carnivore hunt, often followed by a kill, because deep down we are hunters ourselves. I do not think that anyone can resist the lure of watching the incredibly crafty stalk of a cat, the images of the lightning-fast chase by a cheetah, or the long gallop of a dog after a hare. We identify with hunter and hunted, and by merely looking at such predation, whether in the wild or on our TV screens, we satisfy deep urges by proxy.

Finally, I want to illustrate the extensive impact on our culture of these objects of our admiration, and of our anti-predator actions. We celebrate them in literature, in art, in heraldry, in mythology and in witchcraft. Mothers have told stories about the big bad wolf and other predators to their children from early history until today. Artful accounts of such danger come from everywhere around the globe, from African villages to the teeming cities of the modern world.

The instinctive awe and the conflicting emotions associated with carnivores have also invaded our sense of aesthetics, and the images of these animals have become touch-stones. To most of us, the sight of

Polar bear and Arctic fox

a wild leopard is, despite its danger, a breathtakingly beautiful experience, and the silhouette and music of a howling wolf will be forever engraved on the mind of a spectator. Pictures of a polar bear on an ice cliff win prizes in photographic competitions, and the view of a lone fox at the edge of a field somewhere will forever colour the memory of a walk in the countryside. We use the images of these wild animals to describe people, such as a bear of a man, the sinuous, cat-like movements of a girl, or even a foxy politician.

Underlying this appreciation is a biological relationship, between us and them. It needs to be explored, to be understood and admired, and in the end I want our relationship with carnivores to be exploited. This may sound contradictory, but I am seeking to extract every bit of support that we can muster for the conservation and long-term survival of what many of us see as some of the most beautiful creatures on earth. We need them, never mind the fact that some are maneaters and that we are competing with them to secure an existence on our overcrowded planet. If exploitation is the way to sustain their populations, then so be it.

WHAT IS A CARNIVORE?

Carnivore is an ambiguous word, the literal meaning being 'meat eater'. As such, it could describe us people, at least the non-vegetarians amongst us. The word is also sometimes used for predatory animals, even for snakes that kill frogs, or hawks that take sparrows. We and they are all to greater or lesser degree carnivorous. But I suggest that we forget about those more general meanings: there is one group of mammals to which science has actually attached the official label 'Carnivora', and that is the group which claims the title from all others. These are the carnivores that I am writing about: exclusively, the members of the mammalian Order Carnivora.

I will be even more restrictive than that, because I will not be concerned with the seals, sea lions and walrus. These are also often taken to belong to the order Carnivora, as they have evolved from the more terrestrial species, and they are closely related to bears and martens (Bininda-Edmonds *et al.* 1999). However, seals and their relatives have become very specialized and adapted to their aquatic habitat, and the story of their relationship with people is a totally different one. Usually students of ecology and life history consider them quite separately, and also, many taxonomists put the seals and their relatives in a separate Order, the Pinnipedia. Here, therefore, we will recognize the Order

Carnivora in the restrictive sense, next to another Order Pinnipedia (Flynn 1996). In this book it is only the terrestrial carnivores that matter, and it is the members of the Order Carnivora that I will address as carnivores.

Many carnivores are predators, but not all of them. A predator is an animal that kills another one for food, an animal that hunts and preys on others. We will see later that belonging to the Carnivora does not predestine a species to be carnivorous: a panda is a carnivore, despite its diet of bamboo. Nevertheless, our typical image of a carnivore is that of a predator.

Who then are these Carnivora? They may be conspicuous, but compared with other mammalian groups there certainly are not that many of them. Count the numbers of species, or count the number of individuals of each species; whichever way you set about it the score for Carnivora is low. There are well over 8000 species of birds, and fewer than half that number of mammal species, but only 237 of those are carnivores (Bininda-Emonds *et al.* 1999). Moreover, we can state as a generalization that of each carnivore species there are usually fewer individuals around than for most other mammalian orders that live in the same places. Carnivores are often referred to as animals at the top of the feeding pyramid, an image that aptly describes their numerical inferiority.

There may be only relatively few of them, but their effect on others is quite out of proportion to their numbers. Much of their impact is direct, because any effect could hardly be more immediate and final than that of predation. Carnivores kill, and they can extinguish populations. Nevertheless, the *indirect* effects of carnivores may be even more pervasive.

Most animals, whether they are mammals, birds, reptiles, amphibians, fish or invertebrates, are a potential prey for carnivores, and all had to evolve defences, just to protect themselves and their offspring against Carnivora. I will argue later that this affected many aspects of the behaviour and appearance of all land vertebrates (including ourselves) and many of the terrestrial invertebrates. This is evident in such behaviours as foraging, which animals cannot always do with optimal efficiency because of threats from carnivores, or even in mating, which has to happen fast in order to escape predatory intentions at a vulnerable moment. Also, in many other ways animal performance is affected by the need to look over the shoulder, to be aware of predators. Even the colour of many animals is determined at least partly by the need for crypsis, providing protection against predation.

There are spectacular differences between species of Carnivora, but they also have many things in common. To appreciate this contradiction one does not have to be a taxonomist, because for most people there is never any doubt as to whether any one species is a carnivore or not. It may live on a diet of buffalo, beetles or bamboo, but its set of teeth and its overall shape reveal what it is, unmistakably. Not surprisingly, therefore, carnivores are a 'monophyletic' order, i.e. they are species that are presumably descended from a single ancestor. For understanding the human relationship with carnivores this similarity between species is an important point, because our experience with one carnivore is likely to affect our behaviour towards others. If one has escaped an attack from a bear, this is likely to affect future responses not just towards bears, but also towards tigers and wolves.

Over some 54 million years, carnivore evolution produced the present-day rainbow of 237 species from their one ancestor (Bininda-Emonds *et al.* 1999). Their range of sizes alone is telling of their huge variation: species range from a least weasel of 45 g to the brown bear of 700 kg (more than 15 000 times larger), a spread of sizes that is greater than in any other order of mammals, despite the fact that the Order Carnivora is relatively small. Not only are their sizes highly diverse, but shapes also vary between the almost eel-like weasel and the rotund panda. Some species live in groups, others on their own. There are arboreal, swimming, coursing, stalking and digging carnivores, some live in the Arctic, others in tropical rainforests or deserts or the watery depths of rivers, lakes and seas (Macdonald 2001). They are distributed naturally over all continents except for Antarctica and Australia (where some have been introduced, wild or domesticated). There is evidence that this diversity was even greater a million years ago and earlier.

The evolution of this wonderfully diverse order has been particularly well studied, and the phylogenetic relationships of the Carnivora are at the moment probably better understood than those for any other group of mammals. Evidence for their family trees has been collected by many different methods, including various kinds of morphological information from living species and fossils, serum protein, immunological, karyotype and DNA analyses. We now think that the carnivore family tree looks something like Figure 1.1.

Immediately striking in this family tree is, firstly, a large-scale division into four families of dog-like species, and four families of cat-like species. These are two 'clades' that have their origin right at the beginning of carnivore evolution. Interestingly, mankind has taken one classical representative of each of these main groupings into our homes

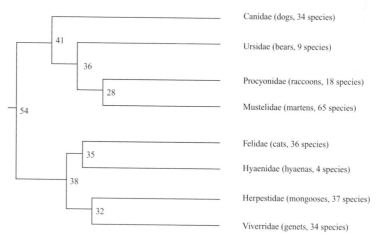

Figure 1.1. Family tree of the carnivores. Dates of branches when the families separated from each other in evolution are given in millions of years before present (Bininda-Emonds et al. 1999).

and domesticated it, and this will be described in Chapter 8. Apart from these main divisions, the family tree also shows, for instance, that dogs are equally close to bears and martens, but cats are closer to hyaenas than to the genets or mongooses. The most recent major family evolution was the split between raccoons and martens, some 28 million years ago.

I should add that there are those who have reservations about the fairly simple family tree as presented in Figure 1.1, and the last word certainly has not been said about it. For instance, there are suggestions that there should be a separate family of skunks, Memphitidae (instead of them being lumped with the Mustelidae) (Dragoo & Honeycutt 1997), and the number of raccoon species is disputed (Pons *et al.* 1999). Presumably, there will always be some variation in the number of carnivore species that are recognized by different authors.

In Chapter 6, I will discuss further details of the evolution of the carnivores, and especially their history in the last few million years, at the time when *Homo sapiens* or its predecessors were also present. Here, I will briefly describe the carnivores in the world of today, in a survey that has to be short from necessity. It is a mere outline of the marvellous richness of this order, giving us some idea of the variety of carnivore predators, of the animals that people admire so much, of the species that threaten us and our livestock, and of what is at stake in conservation.

CARNIVORE GROUPS IN THE MODERN WORLD

The dog family (Canidae)

One of the two best-known families of the carnivore order is that of the Canidae, the dogs and foxes (the other being that of the cats). Their sizes vary between that of a large grey wolf (up to 80 kg) to the tiny fennec fox of the African deserts, weighing in at little more than 1 kg. Canids occur on all continents, and with the dingo they even fielded an early introduction in Australia. The wolf, of course, is the epitome of a canid, spectacular, somewhat threatening, with a beautiful large body and a magnificent sound, and the immediate ancestor of our domestic dog. It is one of the animals we associate with wilderness, and significantly, the image of the wolf also features large in the relationship between carnivores and our own species, as we will see later in this book. Several other canids have featured prominently in my own life. For instance, when I lived in Africa in the Serengeti (Tanzania), I was fascinated to be able to watch three species of jackals around my house, with large packs of up to 40 African wild dogs occasionally passing through when chasing gazelles, and from my window I could often see families of bat-eared foxes catching their termites in the distance.

Red fox

All canids look strikingly dog-like (i.e. wolf-like), even the brightly coloured red fox, slinking along an old stone wall in the pasture landscape close to my home here in Scotland. There are several different foxes on the various continents, the red one being the most ubiquitous, and all are typically canid.

As well as having a large proportion of their looks in common, all species of canids (for example the coyote, jackals, more than a dozen foxes, the African wild dog, the Asian dhole, the South American bush dog, the maned wolf, the raccoon dog) also share basic elements in their ecology and social behaviour. This appears counter-intuitive, because they vary from pair-bonded individuals to gregarious pack animals (see Chapter 2). However, even the pack-living species have an organization derived from a single *pair*: the canid family is the only one where a pair bond is the norm, and where males regularly help with rearing of the offspring. All other families have an organization based on the mother–offspring unit, and males rarely help. Also, canid sounds, as well as their scent marking, visual displays, hunting, prey caching and many other behaviour patterns show striking similarities in all species.

The marten family (Mustelidae)

For some reason I have a special soft spot for the marten family (sometimes called the weasel family) or Mustelidae, and I have spent many years of my life studying them, especially the various badgers, otters and mink. I fell for them after watching badgers in Britain, which toppled one of my prejudices: I thought that I had understood in Africa that predators live in groups in order to cooperate with hunting, but these badgers lived in large clans of non-cooperating individuals, eating earthworms. It just did not fit, and in the process of finding out what was going on I became fascinated by the animals (Kruuk 1989). Subsequently I became attached to otters when I found them sharing a den with badgers along the west coast of Scotland, which led to a long study (Kruuk 1995), and otters brought me in contact with mink, one of my present interests. Those curious, expressionless mustelid faces became an addiction: nothing to do with science, just a bit of an obsession.

The mustelids, with 65 species, are by far the largest family of the Carnivora, and they have shown more recent evolution in numbers of species than any other (Bininda-Emonds *et al.* 1999). They occur in both the New and the Old World, dominating the carnivore scene in numbers of species, and including weasels, martens, mink, polecats, skunks, otters, badgers, wolverine and many others. There are relatively few of them in Africa, perhaps because of competition with the rather similar mongooses found there. In the USA there are more mustelids than other carnivore species, and they include the American badger, wolverine, mink, fisher, marten, black-footed ferret, two otters, three weasels and four skunks (Burt & Grossenheider 1959); in Britain there are also three times as many mustelid species (badger, otter, pine marten, polecat, stoat, weasel) (Macdonald & Barrett 1993) as there are members of all other carnivore families together (fox, wildcat). Mustelids are frequently in the news, because of their alleged attacks on game or poultry, sea otters eating the fishermen's abalones, Eurasian badgers as a hotly contested source of infection for bovine tuberculosis in cattle, otters as a main focus of conservation, animal rights people protesting against fur farms and freeing mink, and in many other roles.

Almost all of the mustelids are quite small animals, the largest being some of the otters and the wolverine. Most are of very slender build, but they can vary from the willowy forms of weasels to a distinctly stocky shape, such as the badgers. They differ greatly amongst themselves in feeding and social behaviour, from the totally solitary wolverine catching reindeer, to martens and weasels stalking rodents, the group-living Eurasian badger feeding on earthworms, and otters, who have evolved as fishing specialists. Unlike that of the canids, the mustelid social system usually consists of independent male and female territories – but there are the odd exceptions, such as the clans

NSF

Striped skunk

of Eurasian badgers, and the complicated sharing of ranges by some of the otters. Males are rarely involved in rearing families. As a rule, mustelids are very silent animals, with few sounds, and also with few visual displays. But scents are very important, with skunks, badgers and polecats being proverbially smelly, and all mustelids have marvellously elaborate scent-marking behaviours.

The bear family (Ursidae)

The nine species of bears and pandas of the family Ursidae are all large, and some are huge. They may weigh in at several hundreds of kilograms, and their size alone would make them threatening to us; in fact, several species are known killers of man. Nevertheless, their looks have given them a public image of being cuddly and friendly: many a child takes a teddy bear to bed, and the giant panda especially has attracted the conservation eye of the world.

The magnificent polar bear is completely carnivorous, but the others are only part-time predators, and for several the majority of their diet is vegetarian (berries, roots, leaves and other succulent parts). Bears are solitary, and one could call them inarticulate, using only a few types of behaviour for communication. But sometimes several individuals will come together at a rich source of food, such as a salmon run or a rubbish tip. Males and females have their own separate territories, and rearing of the cubs is none of the males' business. The Ursidae are a small but spectacular family, and they have successfully colonized areas anywhere between the drift ice in the Arctic and the dense forests of the tropics.

The raccoon family (Procyonidae)

Uniquely amongst the carnivores as a family, the raccoons or Procyonidae are a New World phenomenon. Several of their 18 species are very newly evolved, less than a million years ago (Bininda-Emonds et al. 1999). To me, their image is typified by a single raccoon that stared at me from the water edge when I canoed past in Algonquin in Canada, and backwoods inhabitants will see the ringed tail of a coon as part of their picture of American wilderness. They are small, like the mustelids (the largest ones about the size of a fox), and quite varied in their appearance and behaviour.

Several of the raccoons have a very small geographical range, being marooned on just one or a few islands in Central America, and

Raccoon

there are strong suggestions that these are not proper species but only subspecies (Pons *et al.* 1999) (this still needs confirmation, but it could substantially reduce the number of species in the family). Apart from the raccoons proper there are also three coatis, as well as the kinkajou, cacomistle, ringtail (or miners cat) and several olingos. These species with wild-sounding names inhabit forests and deserts, and raccoons are common even in big North American cities. Their food ranges from meat to insects to plants, with the urban common raccoon scavenging things unmentionable. The social behaviour of the family is equally eclectic: most are solitary but coatis live in large packs. At least, that is what the female coatis with their young do, but males go their own solitary way: at one time people even thought the male ring-tailed coati to be a different species, the coatimundi! The social organization within these packs is highly complicated (Gompper 1996, 1997; Gompper *et al.* 1997). We know less about the social structure of the common raccoon, and there may be considerable surprises awaiting (suggested by the fact that they may winter in groups of up to 23 together, and that they form temporary 'consortships') (Fritzell 2001).

The genet and civet family (Viverridae)

The world's genets and civets belong to the 34-species family of Viverridae, a good Old World family only to be found in Eurasia and Africa. In looks they are probably quite similar to the original ancestors of the carnivores, the Miacidae, although of course we have no idea what colour coat these fossil carnivores had. Many viverrids are the size of domestic cats (1–12 kg, with an exceptional 20 kg for the fossa in Madagascar) with pointed faces, but they are not as highly specialized as cats, with claws that are only semi-retractile. In our home in Africa we often saw the small, spotted figure (complete with beautiful ringed tail) of a wild common genet slip in between our chairs, taking titbits from our hands, and in general people in Africa know the genet as a raider of their hen-coops.

However, apart from the civet cat of perfume-fame (see Chapter 7), the family has few members that spectacularly catch the public eye or nose. They are all solitary nocturnal animals, with few visual or audible communication behaviours, living on a diet of small prey or fruit, often in trees. With these qualifications and because they are largely tropical, it is not surprising that most have only been superficially studied: we know very little about their behaviour and ecology.

The mongoose family (Herpestidae)

All 37 mongooses that make up the family Herpestidae are small (300 g to 5 kg), insect-eating, ground-living animals, closely related to

Common genet and dwarf mongooses

the viverrids, with similarly pointed faces but never spotty. The mongooses too, occur only in the Old World, where they have been very successful in terms of numbers of species, especially in the tropics. In any one area, such as for instance the Serengeti, one can find up to six species: in that case the marsh, white-tailed, Egyptian, slender, banded and dwarf mongoose.

Some of these fascinating little animals, such as the meerkat, and the dwarf and banded mongoose, have evolved a spectacular group organization. Most others are solitary animals, but these gregarious species live in day-active packs of 10–40 members, with a hugely complicated social organization (Clutton-Brock *et al.* 1999). A main function of this pack organization is probably defence against predators (as I noticed when I flew close to a pack of banded mongooses in a small plane on the open Serengeti grasslands: they would bunch up tightly and face me and my plane, like a single multi-headed organism), and they are known, communally and successfully, to rescue pack members from the claws of raptors. The packs also have a communal childcare and guard-duty system (Rood 1986). Solitary mongooses are often nocturnal, and for the same reasons as the viverrids they have been generally very little studied.

The hyaena family (Hyaenidae)

There is no group of carnivores that I find as interesting and sympathetic as the hyaenas, probably just because I have spent a long time

Brown hyaena

studying them. The Hyaenidae family have only four species left, despite a fossil record showing large numbers with at least 61 well-documented species (Werdelin & Solounias 1991) in the geological past. They occur only in the Old World. Three are relatively large animals of 30 kg or more, of which the spotted hyaena, my special favourite, weighs as much as a person. Hyaenas are predominantly carnivorous, although some also take vegetable food, and the small aardwolf is a termite specialist (Kruuk 1975; Mills & Hofer 1998). Of the hyaenas, the spotted hyaena is the most spectacular predator by far, and it hunts large ungulates in packs just like wolves do (see Chapter 2). It has a very elaborate system of communication with displays and highly varied calls, and lives in a complicated and variable system of clans, with the larger females dominating a show of up to 80 clan members at a time. Hyaenas interact with mankind in many ways, and we will meet them in this book in several chapters.

The cat family (Felidae)

Finally there are the predators par excellence, the most carnivorous of all carnivores: the 35 members of the Felidae, the cat family. Animals of the Old World as well as the New, they are all remarkably similar in shape, but with large differences in size and colour. They have penetrated our own world, and whether they be tiger, cougar or domestic cat, we revere them as almost magic hunting machines, with a compelling beauty.

There are small species like the Eurasian and African wildcat, which is almost the prototype felid, the serval and flat-headed cat, the ocelot and many others, ranging up to lynx, puma (cougar) and various leopards, jaguar, cheetah, lion and tiger. The social organization of most species is very similar: solitary and territorial, with males in larger territories than females, just as in most other carnivores. However, some cat species such as the cheetah and feral domestic cats, tolerate a few same-sex adults of their kind in their territory, usually close relatives. The lion is a compulsory group-living exception, with related lionesses living in prides of up to 40 animals, and male lions in coalitions of up to four that live with a pride until expelled by other males (Schaller 1972; Packer *et al.* 1991).

All cats are highly specialized hunters, and they have a large repertoire of visual displays and calls in their social behaviour. Some of the calls are amongst the loudest animal sounds in the wilds: many a human soul has been terrified by the roar of a lion, and hearing it through the canvas of one's tent is an experience few would forget. Many of the felid behaviour patterns show great similarity across species. Partly because one of them has been so thoroughly domesticated and partly because several of the others are so spectacular, our knowledge of the felids is relatively rich, as we will see in some of the following chapters.

VULNERABILITY

Throughout this spectrum of species, in all the families there are many that are now endangered, and whose long-term survival is in doubt. Many carnivores may not yet be extinct overall, or even threatened or vulnerable as a species, but in individual countries or regions they are on the way out or have already disappeared. The black-footed ferret became extinct as a species in the wild but captive-bred ones are being reintroduced in the USA, tigers are seriously threatened with extinction throughout their range, and Ethiopian wolves are only just hanging on, in a few mountain ranges in Ethiopia. The European mink is now almost gone, and of the giant panda only around a thousand are left. In Western Europe wolves, bears and lynx are totally extinct in most countries (although overall these species are not in immediate danger), and in Africa the wild dog is in steep decline and has gone from many regions. One could extend this information into a very long list.

Wildcat

However, and rather surprisingly, the fate of carnivores is not much worse than that of mammals in most other Orders. The Red Data 2000 list (Hilton-Taylor 2000) of the International Union for the Conservation of Nature (IUCN) lists 27.4% of carnivore species as vulnerable, endangered, critically endangered or recently extinct, compared with 53% of the primates, 41% of the even-hoofed ungulates, and 37% of the insectivores. Only orders such as those of the rodents, lagomorphs (hares and rabbits) and bats are less or equally endangered, with 19%, 25% and 27% of species listed. Overall, 27% of all mammal species are in the vulnerable, endangered or critical categories of the Red Data 2000 list, so the threat against carnivores is about average for mammals. This level of danger to survival is especially worrying if one spells out the implications of the different IUCN categories: 'critical' means a 50% risk of extinction in the next 5 years, 'endangered' means a 20% risk of extinction in the next 20 years, and 'vulnerable' implies a 10% risk of extinction in the next 100 years (Mace 1994).

These figures should, of course, be handled with great care and caution. The exact position of a species as vulnerable or critical is to some extent a subjective assessment; there are many species that are 'data deficient' and it is difficult to quantify and compare numbers of endangered species per family. Perhaps some of the mammalian Orders have more critical species, while others have more vulnerable species. However, the data do indicate an order of magnitude of the problem, and this does not appear to be substantially different for the Carnivora.

I was, I admit, rather surprised that carnivores are not in a much worse position than other mammals. After all, they are more often persecuted, as well as being themselves dependent on populations of other animals, and therefore, also exposed to declines in prey species. They must be an inherently vulnerable group, and recently it was shown that species are at especially high risk of extinction when they are at the top of the feeding pyramid, and/or at low population density, and/or with a long period of individual development (Purvis *et al.* 2000a, 2000b). Possibly, carnivores are better able to cope with their extra risks because of greater behavioural flexibility, but this is pure surmise.

Amongst the Carnivora, it is especially the large species that are facing threats, although of course several others are also in trouble. There may be no overall correlation between body size and extinction risk (Purvis *et al.* 2000a, 2000b), but nevertheless of all 17 carnivore species that weigh 50 kg or over, ten (or almost 60%) are on the vulnerable-or-worse list. This includes half the number of bear species. Some families have many species with very localized distributions, e.g.

the raccoon family with several species that occur only on one or two small islands, so they have 44% of species on the vulnerable-or-worse list. Also, as we shall see in Chapter 3, some of the worst affected families, such as the felids (28% vulnerable-or-worse), are exactly those that are the most specialized in their feeding habits. It appears that these are the animals that are least able to put up with persecution, with competition from man, and with changes in their habitats.

However, on a much more cheerful note, there also many carnivores that have adopted man as a lifestyle. They have done well out of agriculture (e.g. coyote, foxes, Eurasian badger, caracal) or they thrive in our urban settings (raccoons, foxes, stone marten, see Chapter 12); or, most spectacularly, they have worked their way into our confidence as pets (Chapter 8).

These simple statements about the plight and thriving of carnivores indicate how people and their fashioned environment have become the most important factors determining the lives of many of the carnivores. Conversely, as we will see later, the carnivores themselves also have a large influence on our existence.

The Order Carnivora may be rather small, but their range of behavioural and ecological permutations is huge and fascinating. We will see in the following two chapters the variety of their social systems – from simple solitary animals to highly structured and large communities, and how this depends on resources which affect these highly efficient predators. The behavioural and ecological background of the carnivores is rich tapestry for the interplay between them and us, which is the subject of later chapters.

Clan of Eurasian badgers and solitary red fox

2
_____

Carnivore society: hermits and communes
Social systems, groups and territories

Behavioural scientists, and probably all of us, are patently biased towards an interest in group-living carnivores, and they get much more attention than the solitary ones. Why is it that people are so taken by the way in which such societies of wild hunting predators are organized? Has it to do with these animals actually hunting in groups, or is it because they organize themselves in group territories and societies, or show other similarities to us? There are fascinating parallels between the societies of group-living carnivores and those of primates. Obviously, one has to be very careful in assigning evolutionary significance to such convergence, but it is at least likely that our interest in highly organized carnivore societies is fuelled by the apparent similarities with our own, or with those of our nearest relatives.

Years ago in the Ngorongoro Crater in Tanzania, I remember meeting up with Jane Goodall, who was taking a break from her work on chimpanzees, which she had recently started at Gombe Stream (Goodall 1971). I had been studying spotted hyaenas for some time there, and we eagerly swapped notes on our animals, discussing individuals and their social set-up. It was difficult not to get overenthusiastic, because there were so many analogies between my hyaenas and Jane's chimps. Most strikingly, both species live in a 'fission–fusion society', i.e. in large communities that have their own territory, but within which the members are not permanently together. Individuals come and go in temporary small groups, they switch between groups or they may stay on their own for a time. It is a social system that is hugely different from that of solitary territories, or from relatively simple, fixed-membership pack living.

That there should be such semblance between species from orders of mammals that are taxonomically quite far apart is fascinating in itself, and in fact any detailed comparison between species is interesting when they are ecologically similar, or related. One is made aware of characteristics that otherwise might be taken for granted, or even overlooked. For instance, Jane Goodall told me about the hunting parties of chimps, which are always males, but hyaena society is dominated by the females, and it is females who take the hunting initiatives. Such small but important similarities and differences between our two subjects were plentiful, and by these comparisons we could start to ask why they occurred, and what the biological function was of such organization.

Spotted hyaenas are said to be the most social species of all carnivores, as they live in the largest groups (which are also very variable in size) and they have highly complex social behaviour (Hofer 1998). We find large societies, 'clans', sometimes involving as many as 80 animals or more (Kruuk 1972a; Holekamp *et al.* 1993; Mills & Hofer 1998). Each spotted hyaena clan occupies a territory (which may occasionally split in two or expand into a vacant area (Holekamp *et al.* 1993)), in which there is a central den, a kind of headquarters. The animals can be excessively social, at times with 20 or more of them literally within touching distance of each other, when they are feeding in a heaving mass. But within this clan system each animal goes its own way: all alone if it is in solitary mode, or ganging up with others at other times. It may go hunting on its own, or get involved in chasing some larger animal with a dozen others. It may sleep alone, or share a lair with other members of the clan. It may help in the defence of the very important territorial border, in a group or on its own, or it may just look after itself. Some hyaenas often go round together, in little cliques; there are highly gregarious individuals, and others that almost always seem to be on their own.

In a commune, cubs are reared only by their own mother. Normally they do not suckle from others, nor do they get provisioned by any hyaena but their own mother (Holekamp & Smale 1990). In contrast, hunting hyaenas share their kill with each other, as well as with other, non-hunting members of their clan, and they all feed together from the same steaming carcass. There is a strict, linear rank order between the females (especially over access to food), and below the females, between the males (Frank 1986). The animals are very vocal, with an enormous range of often very loud sounds for communication (including Africa's most evocative sound, the 'whoop', and you may hear 'giggles', and a large variety of grunts and moans) (East & Hofer 1991). They also have many visual displays.

Does some of this sound familiar? It certainly did to Jane Goodall, who in the chimpanzee was also studying a group-living species with a clear interest in hunting (Goodall 1971). Several of the hyaena-society characteristics apply in exactly the same words to chimpanzees – and for that matter, also to some human tribal communities. In chimpanzees there is a group territory with up to 40 or more individuals; it is defended by singles or groups of animals against neighbours; individuals may sleep alone or with others; there are small 'packs' of (male) hunting chimpanzees, which often cooperate to catch their prey (e.g. monkeys cornered in a tree) and share their food with non-hunters (Stanford 1999); and mothers look after their own young ones. Chimpanzees, like hyaenas, are incredibly noisy, with an enormous vocabulary.

Perhaps the most important aspect that these species have in common is that their societies allow individuals the choice between gregariousness and solitude, on an ad hoc basis. The fission–fusion society may be the most interesting kind of social organization there is, partly because of the possibilities it offers to its members, and because of the high demands it makes upon communication amongst individuals. Understanding it and its evolution is a major challenge that is, of course, a particularly interesting one to ourselves. The fact that this type of society occurs in those primates that hunt regularly (e.g. chimpanzees, humankind), as well as in some carnivores, could be a significant pointer to an evolutionary connection between this type of resource exploitation and social organization. It could mean that these fission–fusion societies evolved for optimal exploitation of combinations of different kinds of prey populations, by either social or solitary hunting.

Apart from the highly flexible fission–fusion organization, many more other social systems have evolved amongst the carnivores, often quite independently from each other in different families. There are the more simple, solitary organizations with little variation, the societies of hermits (this is not a contradiction in terms), where each individual always ploughs its own furrow in its own patch. These are the vast majority of carnivore species. In complete contrast, we find obligatory pack-living species (such as some mongooses, meerkats, female coatis or wild dogs), always within sight or calling range of others, totally dependent on each other for many things in life. Many social systems are more or less in between these extremes, all of them different from each other.

There are only a few carnivores, such as hyaenas, wolves and lions, that live in a fission–fusion community where individuals may cooperate or not, depending on the requirements for the most efficient prey acquisition. It is an ideal set-up for a hunter that exploits large prey that can be shared (and needs help in capturing it), but a hunter that is also dependent on smaller sources of food (to be eaten individually). Other types of fission–fusion societies see animals sharing a den and a territory, even sleeping together, but always foraging on their own, e.g. the kinkajou (Kays & Gittleman 2001) and the Eurasian badger (Kruuk 1989).

The plethora of types of communities is not just overwhelmingly interesting to anyone who studies social behaviour. It also provides a unique window into the processes of evolution, in this case the evolution of social systems. One of the driving forces behind the shape of each organization is food, the result of hunting, gathering and the availability of prey. If we compare some of these carnivore social systems we get a fascinating insight into how hunting may shape the social life of a species. One can ask why a particular organization has evolved that way: what are the environmental selection pressures that favour such a system?

The carnivores show us that cooperative hunting may sometimes be the most efficient way of exploiting particular types of resources, such as large prey animals. But they also show that this cannot be the only reason why animals live in large societies. For instance, several species, like spotted hyaenas, live in clans that are much larger than the groups in which they hunt (clans of up to 80 hyaenas, never all of them hunting together, but hunting packs usually consisting of 1–12 participants). Also, Eurasian badgers and kinkajous live in groups but never forage together. Prey catching collaboration cannot explain this, but there is another factor: the distribution of food. The dispersion of prey animals is a vitally important consideration for a predator's social set-up, as we will see below.

COLLABORATORS IN THE HUNT

We feel intuitively that cooperative hunting is an obvious solution for all sorts of problems, and it comes as a surprise that so many predators do *not* do it. It poses a question that is of particular interest to us, because we are a cooperative species ourselves.

In fact, the pursuit of prey by several individual hunters at the same time is rather rare, occurring in only a few species. These, however, are distributed throughout almost all the carnivore families: amongst stalkers in the felids (lion (Schaller 1972; Heinsohn & Packer 1995, Packer & Pusey 1997), and to a lesser extent cheetah (Caro 1994)), and amongst coursing hunters in the canids (wolf (Mech 1970; Ballard *et al.* 1997), coyote (Bekoff & Wells 1986; Lingle 2000), African wild dog

Banded mongooses

(Creel & Creel 1995), jackal (Moehlman 1986; Loveridge & Macdonald 2001), dhole (Johnsingh 1982; Venkataraman 1995; Venkataraman *et al.* 1995; Karanth & Sunquist 2000)) and hyaenids (spotted hyaena (Kruuk 1972a; Mills 1990; Hofer 1998)), especially when they attack larger prey. Foraging packs occur more rarely amongst grubbing insectivorous procyonids (white-nosed coati (Kaufman 1962; Gompper 1996; Gompper *et al.* 1998)) and mongooses (banded (Rood 1986; De Luca & Ginsberg 2001) and dwarf mongoose (Rasa 1984; Creel & Waser 1994), meerkat (Macdonald 1984; Clutton-Brock *et al.* 2001)) and amongst otters when 'driving' fish (both giant (Carter & Rosas 1997) and smooth otter (Kruuk *et al.* 1994)). The list includes large species as well as small ones.

As one might predict, several of those carnivores that hunt in collaboration do not invariably do so. Wolves, lions, cheetah or hyaenas when in company are more likely to hunt large prey (such as moose, wildebeest, buffalo or zebra), and when on their own small animals (such as hares or gazelle).

The observation that the large majority of carnivores are solitary, with only a few gregarious ones scattered throughout the different families, suggests that most likely, the carnivore ancestor was a solitary species (following the 'parsimony argument'). This means that the fascinating social hunting behaviour must have evolved later, and several times independently in all the carnivore families except the bears. This again means that there must have been distinct and maybe even similar selection pressures operating in these different species.

But curiously, it is not immediately clear *why* some animals hunt socially whilst others do not. There must be benefits, but they are not always obvious, and they may be different for different species. The most heated debate on that point, on the biological function of collaborative hunting, has been over lions, which are probably the best-studied carnivores. It was not surprising that the subject became such a focus: here we have one of the most spectacular animals on earth, showing such remarkable cooperative behaviour that even our own species would have been proud of it, and yet we cannot agree upon its purpose.

The controversy was started by George Schaller, who observed large numbers of lion hunts in the Serengeti, and who demonstrated that lion hunting success increased with the numbers of participating lions, but not as much as one would expect (Schaller 1972). He showed that two lions together had a hunting success almost exactly twice as high as a single one. However, three lions did only a little bit better than two, and with further increase in numbers of lions the numbers of prey caught per hunt remained more or less the same. This at first glance astonishing result means that increasing the number of hunters does not improve success *per hunter*, and there appears to be no immediate benefit of cooperation per hunting lion. Why then do they hunt together?

With my spotted hyaenas I had no such problem, because they catch their prey differently. Two hyaenas attacking a wildebeest mother and her calf were about five times more successful than just a single hyaena, as the wildebeest cow could drive off one attacker but she was powerless against two (Kruuk 1972a). So in this case the immediate benefit of cooperating was obvious. However, this was evident only when hyaenas were hunting wildebeest calves. In their hunts for adult ungulates there was no clear relationship between numbers of hunters and the degree of their success (Kruuk 1972a; Mills 1990).

Perhaps an apt conclusion from this is that our rather simple field observations of hunts do not tell us too much about what actually goes on. For instance, it could be that lions or hyaenas size up a situation, a landscape with prey in it, and adapt their hunting strategy accordingly. If the situation needs a large number, or just one single lioness, then that is what the lions will deploy, and as lions live in a group territory they have the option to do this. This would mean that when we are estimating the success *per hunt* of these animals, we are not, in fact, assessing whether one strategy is better than another in overcoming a prey, but we are measuring something different. It is the animals' ability to decide on the appropriate strategy (including the number of hunters).

A second point is that some hunts will entail much more effort than others, for instance requiring a longer run or a harder fight to bring the prey down. This implies that to quantify the results of cooperation, what we should be comparing is number of prey caught *per unit of effort* of the hunters, not numbers of prey *per hunt* (as was done in the studies mentioned above). It may well be that in a cooperative hunt the participants have to work less hard than in a solitary effort (see below), and in that case collaboration would pay. The issue of the benefits of social hunting is obviously very complicated.

Nevertheless, recent studies have taken us considerably further in our understanding of sociality. For instance, in his beautiful work on lion predation in Namibia, Phil Stander (Stander 1992) demonstrated that collaboration goes much further than several hunters merely aiming for the same quarry. They also take each others' actions into account. Stander showed that in communal hunts each lioness has her own favourite position in the team. There are centre forwards, left flankers and right flankers, and some lionesses are more likely to spook the prey into the jaws of the others. If by chance one lioness ends up in the 'wrong' position, the team is less successful (as any rugby or soccer player would predict).

The team work of coursing carnivores such as African wild dogs is very different from that of stalking predators such as lions. Take this rather typical observation of my own. On an open grassland plain in East Africa, in the late afternoon, a pack of 14 wild dogs trots lightly into the distance. A scattered herd of Thomson's gazelles comes into view, first grazing quietly, then alert towards the approaching predators. The dogs slow down, tightly bunched, heads low and when the gazelles panic, some 300 m away, the dogs start to rush. One takes the lead, closely followed by a second dog, whilst the others seem to spread out rather chaotically. The lead dogs press on as fast and as close to the gazelles as they can. Soon they select one particular buck as their quarry, but when they close in, the gazelle corners with spectacular agility, zigzagging across the grass. The dogs have the advantage, however, because as soon as the buck cuts back one of the following dogs gets close to it, and whichever way it turns, one of the hunters intercepts it. Within five minutes one of the wild dogs, not the leader, grabs the gazelle's flank and bowls it over, and another quarter of an hour later feeding has almost finished.

In an excellent study of these African wild dogs in southern Tanzania, Scott and Nancy Creel (Creel & Creel 1995) showed that the animals obtained most food per hunter for each kilometre they ran when they were in a pack of 13 dogs. When there were fewer or more dogs in a pack, their reward per dog was smaller. The pack size of 13 was optimal in that particular habitat, and it was also the number of dogs per pack that the Creels observed most often.

Studies such as this show cooperation in one of its most spectacular manifestations, and here its efficiency can be demonstrated beyond doubt. But perhaps this is as far as one can go in the analysis of collaborative hunting. Rigorous scientists may demand experiments to demonstrate effectiveness, but any interference with a hunting team would be disruptive in so many ways that it would defeat its own purpose. For the present we may have to be content with the conclusion that cooperation probably helps in certain situations, but in others we do not know yet why it occurs. It is just common sense, however, that for many hunting purposes it will be more effective to hunt alone, when prey is small and not too agile, and when a predator can deal with it efficiently without having to share the catch with others. For most carnivores there are more benefits in being single.

It goes without saying that efficient foraging is not the only evolutionary force that selects for group living and pack hunting. In carnivores this is demonstrated by several members of the raccoon and mongoose families, such as coatis, meerkat, and banded and dwarf mongoose. There the exploitation of populations of (mostly) invertebrates, in broad daylight, is facilitated by a cooperative predator-warning system, by the guarding behaviour of some member of the troop whilst the others forage, The troop members also defend each other against enemies in other ways (Rood 1986; Clutton-Brock *et al.* 1999). In contrast, the Eurasian badger lives in often very large dens, with up to 40 entrances and huge spoil heaps; they, too, live in large societies of sometimes dozens of individuals, but they never go out hunting together, nor do they help each other against enemies, nor even warn each other against danger (Kruuk 1989). They may sleep together in a large ball, but once they leave their den they are on their own. Clearly, the external forces that affect social systems are many and diverse, but aspects of resource exploitation are amongst the most important.

SOCIETY

If hunting and foraging are most often solitary pursuits for many of the carnivores, this may conceivably be one of the reasons that most live permanently alone. Nevertheless, those carnivore societies of solitary species can also be highly intricate and complicated. In the most common state of affairs, the female of any solitary species, be it a grizzly,

Grizzly bear

a wildcat, a marten, a civet or any other, only looks after itself and after its most recently born young. The father has nothing to do with the family, he visits and mates with different females and after that he leaves them to the rearing of his progeny.

Usually, the females have ranges that do not overlap with each other, although they do overlap with the ranges of males. In almost all of these solitary carnivores, the young stay at home in the mother's range until more or less fully grown, but they often leave for pastures new before they are fully proficient at hunting. Their interactions, the contacts between mother and young and their neighbours and various males, all involve many visual and vocal signals, but carnivores especially rely on smell and the secretions of their various glands – and of course such communication is different for each species, as is the timing and the spacing of events such as dispersal.

This is the basic template of the 'solitary' system, and evolution has embroidered variations to it. In the canids (dogs and foxes), for instance, the male range overlaps fairly exactly with that of the female:

the two live in the same range as a *pair*, and they use the same resources, so perhaps we should not even call them solitary. Most important, the males help with rearing of the pups. Canids are almost
the only carnivores that live in pairs (or packs headed by a dominant
pair), and the only carnivores that habitually regurgitate food for their
young. It has been suggested that it was the evolution of this regurgitation behaviour that enabled canid fathers to do their bit for the
family, and that it was this ability that affected the social system of
male participation (Kleiman & Eisenberg 1973), in contrast to almost
all other carnivores. Of course, there may be alternative explanations.

In a further step towards socialization, some of the young stay at
home for a few years into their adulthood, instead of dispersing immediately. In black-backed and golden jackals in the Serengeti, for instance,
young of the previous year help raise pups (Moehlman 1979). They help
with guarding against enemies, they play with the younger cubs and
they help to feed them. The ethologist Patty Moehlman demonstrated
that the more of these 'helpers' were present, the better the pups survived. This does not necessarily show that helpers bring about increased
pup survival, because it could be that better territories or parents cause
helpers to stay on at home. So it may be in the interest of the young
pups, or in that of the helpers themselves, or both. Whatever it turns
out to be, it does demonstrate the beginnings of the evolution of group
societies at their most primitive level. But however basic, such a system
still constitutes a beautifully intricate society, with variations to suit
local conditions.

Even more gregarious, African wild dogs live in extremely tight
packs, sometimes containing 20 or more individuals that are almost
always in very close proximity. On average, a pack contains twice as
many adult males as females, and the males tend to be closely related
to each other but not to the females. The females are usually close relations such as mothers, daughters and sisters. After growing up, groups
of same-sex siblings leave and join up with other such emigrants of
the opposite sex to form a new pack (Fuller *et al.* 1992a, 1992b; Burrows
1995; Woodroffe *et al.* 1997). In principle, therefore, the large social unit
of the African wild dog is an extension of the simple pair on its own
range, as one finds in other canids.

Unusually, however, when wild dogs eat a kill the pups clearly
dominate over the adults. It is amazing to see the little pups, less than
half the size of the adults, come running into a steaming crowd of
feeding dogs, whimpering, and causing all the adults to stand back so

the pups can have the first choice of food. Usually only one female breeds, but all the adults of the pack help with rearing pups (Frame *et al.* 1979). The wild dogs are magnificent to watch, being active by day and extremely tame. One sees the hunters returning from their trip, bouncing up to the den and the small pups running up to them, everybody whimpering loudly, then many of the hunters regurgitate their recently acquired meat for the youngsters. It is animal cooperation at its most exciting, and I will never forget the times I spent with these dogs in the Serengeti, sitting in a vehicle, stationary or fast-moving with the pack, with the animals and their fascinating behaviour within a couple of metres distance.

African wild dogs make an exciting comparison with two other gregarious canids, the wolf and the coyote, which are much more difficult to watch because of their shyness, but just as beautiful. Wolves also sometimes live in large packs, but their fission–fusion type of society sees individuals coming and going, sometimes hunting in groups and sometimes alone, within the pack territory. However, in many parts of its geographical range the wolf is quite solitary, living in pair territories. In general, wolf communities tend to be largest in areas where they feed on large prey, such as moose in North America. Because of such huge variation with environmental conditions, the wolves' social system must be the most differentiated and flexible of those of all the canids.

In the wolf it is both males and females that disperse, the males the furthest, while the females tend to stay close to and sometimes in the range they grew up in. Just as with African wild dogs, usually only one wolf female breeds in the pack, and most or all pack members help with rearing the pups (Mech 1970, 1999). Wolves are much more overtly communicative than wild dogs: they are very vocal and show a vast array of facial and body displays as well as calls. To many people, the howling of wolves must be one of the most beautiful and evocative sounds of the northern wilderness. The wolves' elaborate communication system is likely to be very useful in a society where pack members often leave and come together again, and where a mistaken identity can lead to murder: many wolves are killed by others in territorial disputes.

As far as I am aware, no other species in the entire canid family (as studied so far) has the wolf's versatility of groupings. It enables the animals to deal with prey as small as a mouse or as large as a moose. But the coyote comes pretty close, and it seems almost too much a coincidence that the species looks so similar to the wolf (a bit smaller), and shares much of its North American geographical range. The coyote may live in territorial packs of eight or more, or in pairs, they may catch gophers or pack hunt for deer, or anything in between (Bowen 1982). In the other members of the genus *Canis* we find a range of foods (prey species) and of social permutations that is usually much more restricted. The wolf is obviously equipped to adapt to extremely varied conditions, almost more than any other carnivore: it occurs in North America and throughout Eurasia, from the deep Arctic down into the heat of Mexico and India. Against that background it is perhaps not surprising that the wolf was the earliest, and since then the most successfully, domesticated animal on earth (see Chapter 8). Jointly with the other species of canids, it demonstrates the close connection between ecology and social systems (see 'Resources and territories' below).

The group organizations in other carnivore families show many different patterns – almost as many as there are species. These societies have been described in several single-species books or monographs, for instance on lions (Schaller 1972), dwarf mongooses (Rasa 1984), spotted hyaenas (Kruuk 1972a; Mills 1990), coatis (Kaufman 1962), the Eurasian badgers (Kruuk 1989; Neal & Cheeseman 1996), wolves (Mech 1970) and several others, and in scores of scientific papers. Here again, the scientific record reinforces the impression of our intense preoccupation with the larger animal communities that show so many similar characteristics to our own. Nevertheless, we have to remember that the majority of carnivores are hermits, single and solitary, and that this is the condition from which group organization has evolved separately in different carnivore families.

I found an interesting and more complex society than a mere solitary existence in the Eurasian badger (Kruuk 1989). This species, very common in England, lives in multi-male multi-female groups, at least in areas that are rich in resources such as earthworms or cereals. All group members use the same den, but that is where sociality stops. The system arises as young badgers postpone the day of leaving the parental range for several years (Woodroffe *et al.* 1995). In another mustelid, the Eurasian otter, I found that individuals were living in groups but in an even more solitary fashion than the badgers: several females share and defend a home range, but each has her own 'core area' where she spends most of her time and where she has her den, well away from the dens of other females. Male ranges are quite independent of those of the females (Kruuk & Moorhouse 1991; Kruuk 1995). In the arboreal kinkajou in tropical America, one finds groups that sleep together (usually one female, two males, a cub and a young of the previous year), and like badgers, they forage on their own (Kays & Gittleman 2001).

Some of the most remarkably close societies are the bands of small, insect- or fruit-eating animals. These are the coatis in tropical America, and some African mongooses – the meerkat (Clutton-Brock et al. 1999, 2001), the dwarf (Creel & Waser 1994) and the banded mongoose (De Luca & Ginsberg 2001) – and at times one sees them going around behaving almost like a single animal, tightly packed together. The details of their organizations are highly intricate, and much of their reason for living in packs appears to revolve around defence against predation. They rarely help each other forage (on the contrary, they often squabble over food items), but they warn each other against predators. Meerkats especially have one of the troop as a lookout (Clutton-Brock et al. 1999), and all the social mongooses have warning calls that distinguish between aerial and ground predators. But also, they help with guarding and provisioning the young ones of the other (usually related) pack members, in a structured manner: for instance, female helpers provision most often, and mostly to female cubs, but male helpers provision to both sexes equally (which may be related to females benefiting most from the survival of female cubs) (Brotherton et al. 2001). Curiously, the coatis have a strictly territorial band organization for the females, whilst males are solitary with different food requirements (Gompper et al. 1998).

One could expect that group-living species such as these would be prone to inbreeding. In practice this rarely appears to be a problem. Coati females, for instance, breed mostly with males other than the ones in their own territory. Dwarf mongooses commonly breed with close relatives and do not avoid it, nor does this have a demonstrable effect on their offspring numbers and survival (Keane et al. 1996).

So far, from the comparisons between species there does not appear to be any single striking, simple and general explanation as to why some species are social and others not. Social species are not closely related to each other, nor are there substantial differences in body size, prey size, predation, climate or other factors to explain the occurrence of packs, or bands, or prides. There is also no evidence that social species have done any better or worse than solitary ones in terms of numbers or densities, nor is their survival more or less endangered.

However, although none of the life-history phenomena offer a complete explanation for the occurrence of clans or other groupings, there is a set of environmental factors which is at least quite relevant

African wild dogs

to the issue of gregariousness. These are the factors influencing the distribution and availability of food, or rather that of resources in general. Resource distribution in turn affects predator distribution.

RESOURCES AND TERRITORIES

Carnivore ecologist David Macdonald coined the Resource Dispersion Hypothesis to explain some of the patterns of grouping amongst carnivores (Macdonald 1983). In principle, this idea relates the size of animal territories and the numbers of animals inhabiting each range to the spatial arrangement of resources, and their availability over time. As a standard example that I have already mentioned, Eurasian badgers catch worms and eat grain in cereal fields on their own, solitarily. But despite this, in the rich agricultural lands of north-western Europe they live with many (sometimes 20 or more) individuals together in clan territories. According to the Resource Dispersion Hypothesis, this pattern of grouping occurs because their main food, earthworms as well as cereals, is found in well-spaced patches.

The hypothesis suggests that each patch is available to badgers only over given periods and not at other times, and to be able to feed at all times a badger needs a large area with several patches. It needs to stay in the same area, the territory, because local knowledge of patches is vital. Since each patch carries so much food, each badger can afford to share it with other badgers at any one time. There is a good correlation between the numbers of badgers in each 'clan' and the biomass of earthworms in each territory, and also a good correlation between patch distances from each other and the size of the territory (Kruuk & Parish 1982). Essentially, these badger clans are extended family groups, cramming as many individuals into a territory as can comfortably feed on the patches.

The organization into large groups in other species like wolves, lions or spotted hyaenas is also likely to be affected by the dispersion of the main prey animals. Prey occurs in large, lucrative 'patches', which can be herds, or places where animals can be easily caught, such as waterholes. The most efficient way for predators to exploit these 'patches' is in a system of group territories, for the same reasons as in the territory system of the badger. Then, as a corollary of life in a clan or a pride, and almost as an aside, the predators can start collaborating with each other to catch prey, should that be more efficient. But the fundamental factor that determines the spatial organization of the predator is likely to be the spatial organization of the prey, not cooperation.

This in essence is the Resource Dispersion Hypothesis, and it is a very attractive explanation, although it is far from covering all the intricacies of spatial organization. If this hypothesis is relevant to carnivore organization, resource dispersion could also be an important factor in the evolution of groups amongst, say, primates.

One of the basic assumptions of the Resource Dispersion Hypothesis is that the individual animals or groups live in discrete areas or territories, and in the previous sections I have touched on the concept of territoriality several times. It appears to be highly important for these predators, because it is the one factor common to almost all of them. In nearly all carnivores, at least some (and often all) of the adult animals defend a patch of land against others of the same kind: it is theirs. I have always found territoriality a fascinating behaviour in its own right, and clearly it is of enormous significance in day-to-day existence. It is especially important that territoriality is costly, in terms of energy as well as risk, because it involves fighting and patrolling. On a practical level for conservation management, territoriality is also very relevant for the understanding of ecosystems in which carnivores are involved.

Much has been written about territoriality, especially in birds, where it is very conspicuous. But despite its common occurrence in mammals and birds, we should not take it for granted, because not all animals indulge in it; there is no 'Territorial Imperative' (Ardrey 1966). Our ideas about animal territories probably originated from ornithologists such as Hugh Howard, who as early as 1920 published *Territory in Bird Life* (Howard 1920), a book that was highly influential for a long time. Following the first critical field studies one had the impression that almost all birds and most other animals lived in exclusive areas, in their 'individual space'. When I started my study on spotted hyaenas in the Serengeti in Tanzania in 1964, I was initially quite confused by the fact that individual animals appeared to tolerate each other where I expected them to fight, and I did not see the simple territories that I expected after my earlier work with gulls and other birds. Later it turned out that there were huge, complicated hyaena group territories, but my initial expectation showed how pervasive the idea of a simple system of territories was then.

I had a considerable surprise in Australia, where Menna Jones and David Pemberton, doing their PhD studies in Tasmania on carnivorous marsupials, the dasyurids, showed that they live without territoriality. These Australian ecological equivalents of some of the species in the Order Carnivora demonstrate that hunting animals can organize themselves perfectly well in a non-territorial system (Pemberton 1990; Jones 1995).

The badger-like Tasmanian devil and the various quolls or tiger-cats (which look a bit like martens or mongooses) are to all intents and purposes rather like conventional Carnivora. However, none of them shows any sign of classical carnivore territorial behaviour. The Tasmanian devil is probably the best known carnivorous dasyurid at the moment. It reminds one of a smallish Eurasian badger with large ears, but unlike that highly territorial species of the northern hemisphere, the Tasmanian devil shows randomly overlapping home ranges. A map of these home ranges (from radio-tracking) looks like total chaos, and well over a 100 individuals have been caught (and released) in one site over a couple of years. It is not that these animals are not aggressive: when they meet, for instance over a carcass, there is much fighting with loud screams and hisses. But this aggression is not associated with an area, or with a site, and Tasmanian devils do not patrol any boundaries because there are none. Other marsupial predators, such as the quolls, show the same characteristics.

This Australian conundrum begs the question of why almost all the cats, dogs, otters and other proper (non-marsupial) carnivores (with only the odd exception) spend so much time and effort, and why they risk so much, in the course of territorial maintenance. Why bother with a territory, if the marsupial species can get away without it? There is as yet no satisfactory answer to this: it may have to do with patterns of resource availability, or it may be a phylogenetic difference for which the biological function has gone missing in time. The Australian observations show that the evolution of territoriality is not a simple sine qua non.

The territory defence of carnivores is not just a matter of an occasional scent mark here and there: there is serious aggression involved, and in many species animals are frequently injured or killed during intrusions. One sees threats by visual displays and scent marking, but they would not act as a deterrent unless backed up by the threat of brute force – and brute force it is. I have seen spotted hyaenas being literally torn apart by their neighbours, badgers with terrible injuries on their rumps after territorial intrusion, wolves and lions killed by another pack or pride, wildcats and otters in horrendous fights with fur flying. Man is definitely not the only species that indulges in internecine warfare.

The result of all this aggression is territorial boundaries, which are often rather vague, but sometimes very clearly defined. They are certainly clear in Eurasian badgers, where a border is like a sharp line in the landscape, often a narrow path or a cattle fence (Kruuk 1978a), and similar boundaries are often delineated by spotted hyaenas. But in other species we may recognize borders only on our maps of home ranges and there is nothing to show for it on the ground. In those cases the animals themselves obviously use cues that we do not notice. In low-density populations, or in some species such as the white-nosed coati (Gompper et al. 1998) or cheetah (Caro 1994), the territories may show substantial overlaps, and therefore include disputed areas.

In order to maintain a territory successfully, intruding neighbours should be made aware that they may be attacked by the owner. In almost all species for which the information is available, a male goes for other males, and a female is likely to attack another intruding female. Almost always, the owner is an animal that has invested time and gathered much experience in the territory, and is prepared to risk a fight to safeguard that investment. But often things do not get as far as that, because like many animals, carnivores have evolved a system of signalling to prevent escalation to violence.

A leg lift, anal drag, cheek rub, scratching the ground or a tree, or any other kind of scent mark will later indicate to a visitor that there is a conspecific around. When one visitor meets another animal, then if it is an owner it will smell like its (scent-marked) land and if it is another visitor it will not. Intruders are keenly interested in the scents of the land, and in the scent glands of animals they meet, and it seems that the message is: if you meet someone who smells like the scent marks around, you know who the owner is and you had better retreat (Gosling 1982). Incidentally, I think that this is also the reason why dogs like rolling in smelly substances: it makes them smell of their land. This general, functional explanation of scent marking (by the ecologist Morris Gosling) is much more satisfactory than the common suggestion that a scent mark merely means 'keep out'. The trouble with that last explanation is that any intruder would be able to cheat a message.

There are many different kinds of scent marking: animals use urine, faeces, anal glands, and glands from almost any other part of the body by rubbing against objects or on the ground, or scratching. Individuals themselves are able to recognize other individuals just from their scent, and some (e.g. badgers) can recognize the group an intruder comes from (Kruuk 1989). A scent mark in the landscape provides infor- mation about when an animal passed, how often, and news about its sex and status. There is a highly complicated language out there, but for us human olfactorily blind animals, it is still a largely closed book. Whatever way the signals by non-human animals are transferred, the territorial message is largely the same: I or we have a vested interest in this area, and if we meet, I or we will defend it. The human species behaves quite similarly, but in a more sophisticated manner and with different signals. Nevertheless, the behaviour is still territoriality.

Carnivores have evolved a remarkable profusion of social systems, ranging from solitary, to pairs, or obligate or facultative gregariousness. All of these systems presumably evolved from the solitary state, and all of them are based on costly territorial defence. There is much to be studied still, especially about the environmental pressures that bear on this social variation. The interest in this lies not only in reaching an understanding of evolution, but of territoriality, and of the tendency to hunt in packs, or the habit of living in group and fission–fusion societies with or without the benefits of cooperation, all such phenomena of our own behaviour as can surely be better understood if we know more of their ecological correlates in other, comparable species. It can only help our knowledge of ourselves if we take an interest in the ways in which hunters organize their communities, ways that are certainly different from ours, but not so different that we cannot see reflections of our own world.

Leopard with kill

3

The quarry of the hunter
Carnivore diet and hunting behaviour

The participation in a stalk, in a long chase, in the battle with a victim and seeing its hairbreadth escape, all this must always have been one of the great excitements that life has to offer, right from the beginning of our existence as a species. We may experience this at first hand, as hunters ourselves, or by proxy in front of the TV. But even those who themselves are not hunters in some way or other, and people who disapprove of hunting as an activity for *Homo sapiens*, cannot help but be keenly interested in (although perhaps horrified by) the drama of a hunt and its outcome. Carnivores have an unsurpassed expertise as hunters, which is one of the big attractions of the animals for us: they demonstrate the ultimate skill in deciding over the lives of others. The behaviour of a carnivore, the 'life of the hunter', is a subject of endless fascination, sometimes even envy.

Part of this allurement may be due to us identifying with either the prey or the hunter. Perhaps we feel instinctively that what we observe of the hunting process may be useful some day, as it would have been in the early stages of human evolution. How do these role models, potential predators on ourselves and competitors, relate to their environment, especially to their potential prey animals?

DIET

The large majority of the Carnivora species eat meat to a greater or lesser extent, and they will prey on other animals at some time or other. But as the ecologist John Gittleman pointed out (Gittleman 1985), many of them also take other foods, and some are totally vegetarian. If you call a species carnivorous when more than 60% of its diet consists of mammals or birds, then only 36% of all Carnivora qualify. The definition is, of course, quite arbitrary. In another study it was found that, of the large carnivores, i.e. 20 kg or more in weight, more than half feed entirely on vertebrates (Carbone *et al.* 1999). Of the smaller carnivores, only a quarter take just vertebrates for prey, and the rest feed on a mixture of vegetables, and invertebrate and vertebrate prey.

We know a great deal about carnivore diet, food selection, feeding and foraging behaviour. In fact, we probably know more about these aspects in carnivores than we do in many other mammals, because diet analysis in carnivores is relatively easy. Food usually consists of clearly separate items, which can be easily recognized and quantified in faeces or by direct observations (unlike most vegetation). Furthermore, carnivore scats (faeces) are often easy to find, however elusive the animals themselves may be. Scat analysis has become a major tool for ecologists, despite the problem of quantitative interpretation (Carss & Elston 1996).

It is also often possible to get good direct and distinct observations of predation, i.e. the process of catching animals for food. This contrasts with, for example, grazing, where it is much more difficult to see exactly what is going on. When I watch an otter in the loch near my house in Scotland, I can often see what kind of fish it brings to the surface, whether it is an eel or a perch, and I can estimate its size. A pride of lions killing a zebra or a pack of hyaenas killing a wildebeest provides easy data points, and so does a stoat jumping on a rabbit. The result of such observations of predation, and of the hundreds of thousands of scats that have been analysed, is an extensive body of knowledge of diet as well as of hunting behaviour. This knowledge is an important link in our understanding of the relation between carnivores and their environment, and it has generated many hundreds of scientific publications.

When discussing food it is tempting to go into details for every predator. However, I think instead that it is more useful here to make some generalizations, without forgetting that every species has its own agenda. For instance, when considering the subject of diet (e.g. whether vegetarian, or eating small or large vertebrates, or insects) it is striking how important the taxonomic position of each predator is. The dietary pattern of one cat species is more like that of another cat than that of a dog, and so on. This is especially true when we compare species belonging to the same genus, but it also applies to families (Clutton-Brock & Harvey 1984). By and large, each family has its own dietary pattern, and within that family each species shows variations on the family theme (Kruuk 1986).

The most typical, meat-eating carnivores are the cat family, followed by the mustelids. Bears and pandas are mostly vegetarians (with one exception, the polar bear). The viverrids, the civets and genets (which are most like the 'original carnivore'), are all either insectivores or have a very mixed diet, and the same is true for the closely related family of the mongooses. The diets of the four hyaena species include many more food categories than one would expect of such a small group, with their specializations ranging from wildebeest to termites, from melons to carrion. The raccoon family all have a mixed diet, containing a lot of plant food. The canids have varied, but mostly carnivorous diets, consisting of mammals, insects and fruits. Many of their species are predominantly meat eaters – but even those usually include some vegetable matter in their diet, quite unlike, for instance, the cats (summary in Gittleman 1989; Kruuk 1986).

These family-specific trends are further refined in the subfamilies, where the similarities between species are even greater. For instance, the large and mostly meat-hunting family of mustelids also includes a subfamily of nine badgers, which almost all feed on invertebrates and vegetation (Neal & Cheeseman 1996), and there is another subfamily of 13 otters, which subsist on fish, frogs or crabs (Kruuk 1995).

Not only are there differences between families in the kinds of prey or plants they select, they also vary in the *degree* of specialization. This is important, because specialization per se can affect an animal's vulnerability to environmental change. One can quantify specialization from the number of major food categories that an animal consumes. For instance, in one study (Kruuk 1986) I used four categories of mammals (small rodent-sized, rabbit-sized, larger up to 50 kg, and mammals over 50 kg), as well as carrion, fish, plants, invertebrates, birds, and amphibians/reptiles. Of these ten food categories, I found that amongst the canids a species uses an average of 6.5, each constituting at least 1% of its diet. This means that most canids are not very specialized.

At the other end of the scale, each bear species only uses an average of 3.7 food categories, and each felid 4.0, so they are more specialized than the average canid. The other carnivore families are in an intermediate position on the scale. High specialization implies dependence on few resources, so a specialist has less to fall back on when these resources get squeezed.

One problem in such comparisons is that one can describe specialization only in the broadest of terms, because actual food selection is difficult to quantify in the face of differences in food availability, which have to be assessed. We often use words such as omnivore, opportunist,

Grizzly bear with salmon

generalist and specialist. But these labels have no absolute values, and they may mean different things to different people. An omnivore, even in the Serengeti, might not eat elephants or fish. More complicated still, the Eurasian badger, for instance, may be highly focused in its food selection in any one area, concentrating on earthworms in north-west Europe (Kruuk 1989), on rabbits in southern Spain (Martin *et al.* 1995) and on olives in northern Italy (Kruuk & de Kock 1981). There is no doubt that in each of these areas badgers are highly special-ized compared with the other predators around. Nevertheless, their specializations are different in different places. There is still earnest scientific debate about whether this animal is an omnivore or a spe-cialist (I call it a local specialist). This is, of course, merely a matter of terminology, not of data, because we know a lot about what these animals eat.

However, despite inadequate terminology, we can recognize that some species rely on many more different prey categories than others. For instance, a cheetah on an African savannah kills almost only an-telopes out on open grassland (Caro 1994), but a similarly sized leopard in the same area is much more catholic in its tastes. It eats those same antelopes, but it also takes smaller mammals, and birds, snakes and carrion (Bailey 1993). It takes them in the open as well as in dense bush or between rocks. Similarly, along European streams a mink will eat small mammals, frogs, fish, birds and insects, but along those very same banks an otter will feed almost only on fish and frogs (Sidorovich *et al.* 1998). In all such situations one species is clearly much more focused than the other, and, therefore, we have a broad comparative indicator of specialization. The terminology serves to suggest that a particular predator may be dependent upon few or many prey categories in any one area, and that distinction is useful.

So we have seen that members of families or subfamilies have aspects of their diet in common, and that they often look similar and act alike. Of course, there are many exceptions to this. For instance, the polar bear is the glaring exception in the entire bear family, as it is the only entirely carnivorous bear amongst congeners that are largely vegetarians. The hyaena family consists of the aardwolf feeding on nothing but termites, the spotted hyaena as an exclusive large ungulate hunter or scavenger, and the striped and brown hyaenas with tastes as catholic as one could expect (Kruuk 1975). But these exceptions do not invalidate the finding that ancestry often defines an animals' environmental relationships – such as feeding ecology (Clutton-Brock & Harvey 1984). I will return to this later, when we discuss the way in which we generalize our own reactions and behaviour to predators.

One other rule of diet is quite obvious and relates to animal size: the larger the predator, the larger its prey. This has been generally accepted for some time (Bourliere 1963), and it makes intuitive sense. The small carnivores in particular exploit the huge biomass of insects, earthworms and other invertebrates (as well as vertebrates), but ecologists have shown that the large predators, the ones over about 20 kg in weight, are simply forced to use larger prey. They just cannot feed fast enough to subsist on invertebrates, despite the fact that there are masses of such potential prey available (Carbone *et al.* 1999). Therefore, we see that amongst related large carnivores, overall prey size increases commensurate with the size of the predator. However, there are complications (Kruuk 1986).

The size rule works well within a family such as that of the cats, with the small wildcat feeding on grasshoppers and voles, the lynx on hares, the cheetah on gazelles, the lion on zebra and the other species fitting neatly in between. Similarly, the size rule works within the canids (summaries in Gittleman 1985; Kruuk 1986; Estes 1991). But in other families, such as the mustelids, there is no significant correlation

Lynx

between predator body size and prey size. The tiny weasels and stoat take rabbits and voles, whilst the much larger Eurasian badger feeds on earthworms. Members of the family of the largest of all carnivores, the bears, are perhaps surprisingly mostly vegetarian, although they also often take animal food. Many of the extremes in size, be they very small or very large, are generally highly specialized species such as bears, tiger, weasels and dwarf mongoose.

As predators on other vertebrates at some time or other (even when not most of the time), many carnivores have been and could still be potential competitors of our own species, of the hunter and the farmer; often or at least occasionally, the larger carnivores in particular may take very big vertebrate prey, which is also the main focus of human hunting and livestock interest. Furthermore, people themselves are potential prey. Human-sized prey forms part of the diet of the larger canids, of the bears, of spotted hyaenas and of the large cats. Conflict between man and carnivore, therefore, is almost inevitable, because of dietary specializations and vulnerability to predation. Both carnivore and mankind may play the role of competitor, of predator or of prey in this game.

HUNTING BEHAVIOUR

The composition of the carnivore diet is directly relevant to the interaction between the animals and humans, and, as we shall see, it provides part of the ecological background for our own behaviour in relation to carnivores. As a first requirement, we know from studies of carnivore diet what are the *kinds* of species that are vulnerable to predation. Diet is the result of foraging and hunting behaviour, and it is the outcome of the predators' interactions with prey. We therefore need an understanding of hunting behaviour, in order to describe the effects of carnivores on prey individuals and populations, and to help to understand the animals' significance to mankind.

There are many more hunting strategies used by the carnivores than there are species of hunters; no two hunts are ever the same. Yet it is possible to discern common patterns, and to see that all hunts are derived from one template. Here, I want to concentrate on these common denominators, by firstly describing some observations of ostensibly very different hunting strategies, and then analysing and describing what exactly they have in common. I start with an observation of my own.

I am scanning the open, sunlit Serengeti grassland, with its scattered acacia trees and a few small shrubs. A cheetah walks quietly, head up, alert. Here and there a few Thomson's gazelle are grazing, unaware, and ignored by the cheetah. After 20 minutes the animal stops. It stares intently then walks slowly, with head held low, towards three female gazelles some 200 m away where clumps of tall grass provide cover. Several times it freezes when one of the three lifts her head. About 100 m from the gazelles the predator gathers speed, running with long, rapid strides, and is detected by the quarry at only 60 m distance. A fast chase follows, but this is a matter of only seconds. The cheetah closes in on one of the gazelles, and at full speed slaps the victim's hind leg with a front paw. It was the gazelle's last run. She somersaults, then the cheetah is above her and bites her throat. Two minutes later the kicking stops, and after one more minute the cheetah lets go, panting, and waits for a few moments before starting to eat.

A contrast to this is a hunt by a different predator on another continent, watched by wolf ecologist David Mech from the air on Isle Royale in Lake Superior, Michigan (USA), in February 1960 (Mech 1970).

A pack of 16 wolves travels along the shore of the island, in the middle of the day. Suddenly they veer inland, where 200 m away a moose cow stands on a ridge. The cow starts running when the predators are still 100 m off, and the wolves charge in hot pursuit. There is no attempt to stalk or hide, just a straight galloping chase. The hunters soon catch up, and the moose stops next to a spruce protecting her back whilst she faces the wolves. The wolves lunge and the moose kicks frantically with all four legs; they clash for about three minutes. Suddenly the moose loses her nerve and runs, with the wolves biting at her flanks but releasing again. The whole party plunges down a very steep slope, and when the cow comes to a halt at the bottom the wolves are all over her, hanging on to her back, flanks and nose. The cow goes down, and after a struggle lasting for ten minutes the moose expires. The wolves start eating well before she dies.

We can also compare these observations with the behaviour of a Eurasian badger that I watched, in a woodland in the south of England in May, close to midnight. I am on foot, following a radio-collared animal. In the moonlight I can just see the badger with my star-scope, a night-vision device. The animal quietly trundles through the shadows before reaching the edge of the woods, squeezing under a fence into a large, short-grass pasture. It slowly noses across the grass, zigzagging, stopping and starting. It stops again, then with a quick head movement it grabs a large earthworm. The little delicacy is some 20 cm long and was lying on the surface, with its tail in its burrow. The badger snaps up the worm, whipping it out of its burrow, and I can hear champing teeth. The whole process takes 10 or 15 seconds, then the badger moves on again, very quietly, landing another worm 20 seconds later. It continues to spend almost 3 hours 'worming', exploiting an extremely abundant resource in only a very small patch of its range.

Each hunt of the cheetah or a pack of wolves, each foraging trip of a badger is different from the next one by the same animals. Nevertheless, observations such as the above three are typical for the species (Kruuk 1978b; Caro 1994), and hunts like these contain all the important elements of foraging behaviour in any one carnivore. If we generalize, we see a pattern such as this: before any contact with prey is made there is first the *search* (a behaviour strongly affected by hunger), which then leads to detection and selection of a potential quarry or of food. It is followed by the *approach*, which may involve stalking (e.g. by cats) and/or chasing (e.g. by canids). The actual *capture* may include grabbing, immobilizing and killing, which in some species such as cats and mustelids is done by a dedicated killing bite to the neck. This is finally followed by the successful predator *eating* the catch, or taking it to cubs, or sometimes quietly caching it for later consumption.

What I have described here is a complete carnivore hunting sequence that leads to feeding, but some parts or other are almost always absent. Most carnivores (like the badger above) will just search, then eat small food items without much further ado, almost like a herbivore does. But even if there is a full-blooded hunt, some species never stalk, and others never chase. Furthermore, many carnivores may not show any specially directed killing behaviour after they capture a prey, but just eat it alive.

In this way, every hunt is a variation on a species-specific theme, highly adaptable to the requirement of every case. For instance, lions are fabulous hunters, the ultimate stalking and killing machines. But one large male lion that I saw asleep and then waking up in the Ngorongoro Crater in Tanzania only had to turn around to claim a wildebeest, conveniently killed by a pack of spotted hyaenas right next to him. All the hassle of search, approach, capture and kill had been removed. Similarly, a fox in a hen-house only needs to capture and kill, as there is no need to run and chase. It means that the sequence of events that is called hunting varies greatly according to circumstances: it depends on both the prey and the environment, and the predator's own motivation and its degree of hunger appears to affect especially the early, searching stages of the train of events. If a prey is easily available in front of its nose, then a predator does not need to be hungry but it will readily take and kill the quarry.

Are some kind of hunts more successful than others? It seems a silly question to ask: of course some of the predators' exploits will succeed, and others will fail. But measuring and comparing success in different species and in different situations is difficult, and that is one reason why in the previous chapter, we could not always assess whether cooperation in hunting is really worthwhile.

Let me explain with another example, by comparing the hunting of cats and dogs. Even if each catches an item of prey in, say, 30% of their hunts, success may still be very dissimilar. A cat may catch small mice and a dog large rabbits, but a cat may have moved a few metres over hours and a dog kilometres in minutes. In other words, it may not be very illuminating to compare a cat hunt with a dog hunt, but we should compare the amount of energy gained with the energy expended in each of the two species. For most carnivores we are still a long way from achieving this, because energetic measurements in the field are difficult and expensive. Where energetics have been studied the results have been very interesting. For instance, in both otters and African wild dogs the predators were found to spend an enormous, unusually large effort in hunting, expending a very high amount of energy (compared with their total energy turnover) (Kruuk 1995; Gorman *et al.* 1998). This energetic investment then resulted in a large prize, a large quantity of high-quality food. But clearly these species are gambling with a high-risk strategy: if, in any one area, the returns for their efforts should decline by, say, 20% or 30%, they would be unable to sustain their efforts. When we watch otters or wild dogs in the field we are amazed by their apparent success – but at what a risk.

I have mentioned the problem of comparing success for different social formations in hunting: solitary or in groups. The success of such hunting may be unrelated to group size, and this can be explained by the assumption that carnivores only set out on a hunt in a particular manner (e.g. pack size) or situation if they perceive that there will be a given chance of success. In that case, the success that we measure is the ability of animals to gauge their chances, and not necessarily the efficiency of a given hunting method. Because of these methodological difficulties, we cannot yet compare the efficiency of the hunting behaviour of different species or populations, or in different habitats. This is a serious problem, because hunting success in these contexts is directly relevant to the understanding of carnivore evolution.

Phylogeny, the ancestry from which animals have descended, is, as we have seen, important in determining diet. It is also relevant when we clarify how different animals set about catching their prey: mention your relatives and I will tell you how you hunt. We can make some broad generalizations: at the extremes, cats are the ultimate stalkers and ambushers. Many of the martens also stalk, but dogs and hyaenas hardly show this behaviour at all, being coursing predators that run after their prey. Most of the bears, raccoons, genets, mongooses and skunks are foragers that grub around in the undergrowth, whereas badgers are diggers and otters swim and dive.

SURPLUS KILLING AND CACHING

There is one spectacular phenomenon about carnivore hunting that is the cause of massive human opprobrium. It is surplus killing, or slaughtering more than is required for immediate consumption (Kruuk 1972b).

When a fox or a mink or a genet or a raccoon gets into in a henhouse, it may kill all the occupants, and eat only one or two of them. I found foxes after an orgy of killing over 200 birds in one gull colony. Surplus killing is known from lions amongst herds of cattle, hyaenas killing a whole herd of gazelles, polar bears surplus killing narwhals, wolves killing many caribou, leopards killing dozens of goats, or mink slaughtering scores of terns on their nests. In all these incidents large numbers of animals were slaughtered without being eaten. Presumably all carnivores, whether small or large, are capable of surplus killing.

Fox

What these situations have in common is a lack of defence by the prey, for one reason or another. The prey may be immobilized by particular weather conditions, or it may be penned in, or it may have lost its anti-predator defence through domestication. The predator may be sated and no longer hungry, but as we saw above, hunger normally motivates only the early stages of the hunting sequence and especially the search. Sated with food, an animal will not search for more. But if no search or stalk or chase is needed, because for whatever reason the hunter suddenly finds itself close to the quarry, then the rest of the hunting sequence is put in train and the killing starts. The predators have no specific inhibition to stop the killing if there are many prey (apart from fatigue, perhaps).

Surplus killing is wasteful not only in our eyes, but also from the carnivore's point of view. It reduces prey availability and the predator does not benefit. Why then, do the animals do it? The explanation, I think, is that what we see are the cases where the 'normal' course of events has gone awry. Usually, the behaviour design of predators serves them well, and they should kill whenever a chance presents itself, because such occasions are few and far between; even if the killing orgy results in a massive surplus of food, and even if the predator cannot consume all of its apparently wanton slaughter immediately, all is not lost.

There is one kind of behaviour that appears to be designed to limit the waste. It is food caching, the hiding of a prey for later use. Many carnivores do this (Vander Wal 1990), and in many different ways. I will give just a few examples. Caching is well known as a highly stereotyped behaviour in all the canids, and it is also seen in our domestic animals. Typically, a pet dog or a fox or a wolf searches for a quiet, hidden spot, digs a small hole with its forefeet whilst holding the piece of food in the mouth, drops the item in the hole and then sweeps dirt over it with its snout.

All and only canids cache food in this particular way (so most likely, they all inherited it from the canid ancestor), and in the other carnivore families there are many variations on the caching theme. In the Serengeti, I have seen leopards take carcasses high into a tree, and spotted hyaenas cache chunks of food in shallow water (amazingly, they can find the food again later, merely by repeatedly dipping their head in the murky water in about the right spot). Brown and striped hyaenas push a wildebeest leg out of sight into a dense bush, and brown hyaenas may hide many ostrich eggs this way. Stoats, and some other mustelids (especially mink), sometimes make large stores by dragging different prey items into a single, existing hole. Wildcats may put remains of their quarry under a log, and pumas may scrape leaves and branches over a carcass. Polecats store live frogs in their dens, after immobilizing them by biting their heads.

In some carnivores, food caching has evolved to a fine art. Foxes are able to remember where they stored what, and they return preferentially to the more desirable cached items, sometimes months later. A tame fox, taken for walks around the countryside by David Macdonald, first emptied its caches that contained the favourite meadow voles, and returned only later to those of the less palatable bank voles, scattered in between the other ones (Macdonald 1976). In order to add a sophisticated touch, foxes also use some kind of bookkeeping system for their caches, leaving a drop of smelly, long-lasting urine near those food stores that they have emptied (Henry 1977).

Caching must have evolved in carnivores on several different occasions. It is a fascinating adaptation to surplus food, but even that still does not utilize all the apparent waste. Not only are some of the caches never revisited by the owner, but from the larger surplus kills the remains of only a small number of the victims are stored by the predator. But nothing is perfect, and at least caching does reduce the waste.

Hunting itself is far from infallible, and many a time does the prey escape. But overall, the ability of carnivores to penetrate the defences of other animals, of many birds and all other land mammals, and to use them as their main resource is a wonderfully adapted set of behaviour patterns. It is something that will never cease to impress mankind, which is also a hunting species. With the potential of such adaptation, it really is not surprising that carnivores have also managed to exploit people as prey, vulnerable as we used to be and often still are.

Stalking tiger in the Sundarbans

4

Man the hunted
Maneaters

Carnivores show a strength in diversity, with their fascinating social systems and different hunting behaviours. Numerically, however, they are weak, compared with other animals. Nevertheless, despite their small numbers, they have an effect on people that is out of all proportion to their abundance. In order to analyse this in some detail, I will start at the darkest, most horrifying and negative side of our relationship with the animals, that is their predation on us. They can be very dangerous enemies.

For obvious reasons conservationists often deny that large predatory animals actually kill people, but there is ample evidence that such indignant denial is nonsense. We will see in this chapter that there are considerable numbers of carnivores that actually prey on us. The details of such predation are often anecdotal, and I will present them as such, but these occurrences are nevertheless real, and as my former teacher Niko Tinbergen used to say, many anecdotes make a statistic.

The stories are bloody, and some readers may be put off by the gory detail. Such a reaction is part of our anti-predator behaviour. But I think that the pattern of predation is important, as is how common the incidents are, because this is what makes up the threat which, in evolution, has shaped our response to predators. Amongst other things, we want to know whether any group of carnivores is more of a threat to us than others and if so, how these animals operate.

Our anti-predator response may be 'instinctive', i.e. more or less hard-wired into our brains, as it is in many animal species. In addition, or alternatively, it may be passed on to us by others in our culture, perhaps even by parents reading bedtime stories to children. Whatever is the case, nature or nurture, to assess an ecological basis for the human response to carnivores we need, first of all, an objective analysis of predation on our species. We need data to describe predation on people, from direct evidence.

There is a substantial body of good information, from many species of wild carnivores all over the world, with blow-by-blow descriptions of attacks on people. I have reviewed data for individual species of carnivores from this rather gruelling set of accounts, and at the end of this chapter I have tried to derive some general conclusions.

One of the problems to be assessed is the reliability of accounts, because such information often does not come from peer-reviewed scientific papers. The criteria I have used are that sources should either have been evaluated by other, qualified reputable scientists (e.g. authors of well-known books) or that reports should come from local and authoritative sources (e.g. community registers). The names of sources are given.

TIGER

In the early 1970s Hubert Hendrichs, a zoologist whom I had previously known as a colleague in the Serengeti and is now a professor in a German university, did an extensive study of the remaining population of tigers in Bangladesh (Hendrichs 1975). Hendrichs is a large man, and his size is relevant, as we will see below. He concentrated on the problem of maneating, and the following is one of his reconstructions from tracks in the mud and evidence from survivors:

> The place is Mara Passer, in the huge mangrove forests of the Sundarbans, Bangladesh: February 10, 1971, about 10 a.m. A team of woodcutters is working in a forest clearing. A large male tiger approaches from the southwest, very slowly. A few metres from the clearing the animal crouches, absolutely still, totally invisible under its eerily beautiful camouflage. It assesses the open space, the people and the distances involved. The tiger does not wait long before launching into a fast sprint, directly towards the nearest man, jumping on him from about 3 metres. The woodcutter, a strong, thirty-year old, is unaware of anything amiss until he is hit from behind.
>
> The tiger grabs the victim and almost bounces back together with the man, pushing himself off from a tree trunk without touching the ground. The attacker claws the man's back and shoulder, and holds on to the right side of the victim's neck with its enormous canines. The tiger drags his prey into the forest, going south along a small deer path. Fifty

metres on it drops the victim, disturbed by the other woodcutters. The dead man is found on his back, with his eyes wide open in frozen horror. His knees are bent, his hands point up towards the sky, and there is a pool of blood under his neck from the deep tooth marks.

On that occasion, the body of the unfortunate victim was taken away by his friends before the tiger had the opportunity of eating from it. But four days later, the same animal killed again. This time the body of the man was left on the ground by his companions, and a shikaree (game scout) waited on a machan, a wooden platform in a nearby tree. The tiger was shot when it returned to eat the corpse, bringing to an end a well-documented series of killings. This one predator had killed 32 people in the same general area, 23 in the last two months before its death, at intervals of up to ten days. Twice it killed two people in one day.

Hendrichs cannot convey the sheer horror of the incidents in his hard figures, or even in his haunting photographs. Nor is he talking about a one-off occurrence or about a single, occasional tiger that has gone astray. He collected data for the Bangladeshi part of the Sundarbans, an area of about 6000 km^2 out of a total 10000 km^2 (the other part of the Sundarbans is in India). Two-thirds of the area consists of mangrove forest and the rest is water, with huge tidal rivers and streams. In the Bangladeshi part a total of 392 people were killed by tigers between 1956 and 1970, an annual average of 26, or 0.6 deaths per 100 km^2 per year. In some areas the killing rate was as high as 1.8 people per 100 km^2 per year. To see these figures in perspective: few people live in the Sundarban forest reserve, but about two million people surround the area on the Bangladeshi side, and they use the reserve area extensively.

The Sundarbans maneating tigers are usually males. This may be related to body size, because male tigers are considerably heavier than females, with normal weights being between 130 kg and 160 kg, two or three times as heavy as their human prey.

It is alleged, at least in the Sundarbans, that people are suboptimal prey for these predators, and many of the maneating tigers shot were in rather bad condition. Hubert Hendrichs suggests that tigers would not be able to sustain themselves on a diet of just human beings. They also need to take deer, but those are relatively plentiful in the dense forest areas.

There are strong rumours that maneaters are selective: for instance that when a tiger takes a person from a group of people in a boat, it often selects the fattest. In the last century the traveller Constable, translating the Frenchman Bernier who visited the area in the seventeenth century, tells us:

> among these islands it is in many places dangerous to land, and great care must be had that the boat, which during the night is fastened to a tree, be kept at some distance from the shore, for it constantly happens that some person or another falls prey to tigers. These ferocious animals are very apt, it is said, to enter into the boat itself, while the people are asleep, and to carry away some victim, who, if we are to believe the boatmen of the country, generally happens to be the stoutest and fattest of the party. (Constable 1891)

Hendrichs relates with a certain amount of self-interest that the locals are still convinced that large, fat people run a greater risk (although there are no data to support this).

Tiger attacks occur throughout South-East Asia. In January 1998 a tiger attacked and wounded two people in Khao Yai National Park in Thailand (the animal was shot afterwards). Twenty years earlier in the same area, a female tiger killed a young girl, just below her house, as she tried to retrieve a pencil that she had dropped from a window. Later a game ranger sat up next to the window, two metres above the ground, to try and shoot the animal when it returned. But when he poked his head outside the tiger jumped at him and mauled him, and he died the next day (Khao Yai National Parks Authorities, pers. comm.).

The legendary British hunter Jim Corbett lived in India during the first few decades of the twentieth century; India's finest national park is named after him. He compiled dozens of cases of maneating by tigers and leopards, and some of his statistics are nightmarish (Corbett 1991). Of the eight maneating tigers that he documented (three males, five females) one female had killed 436 people, and two others accounted for 64 and about 150 villagers. Each tiger preyed on people over a long time, 5 years or more, before it was shot. Corbett describes the deaths of 34 people in detail. All the victims were killed in the day-time when cutting grass, collecting firewood or tending cattle, with men and women seemingly equally vulnerable (16 were women, 18 men). Such accounts are presented in such detail that they convey a very factual picture. One feels that Corbett took great pains over the truth.

George Schaller, the well-known American naturalist and writer, also uncovered some statistics in India, which he clearly endorses (Schaller 1967). He found that one district alone averaged 200 to 300 people killed by tigers per year during the nineteenth century. Official government statistics for the whole of India showed that in 1902 as many as 1046 people were killed by tigers, and one can be pretty sure that only a fraction of all kills were reported. Elsewhere, scientists reported a minimum of 189 tiger attacks on people between 1979 and 1984 (Kothari *et al.* 1989). But as Schaller writes, maneaters have become rare in most areas nowadays, because today the cats are shot before their depredations on man become really extensive.

Other protective action against tiger attacks is also alleged to have been successful. In areas where maneaters are active, electrified dummies of people have been deployed. Furthermore, it was noticed that tigers usually attacked people from behind, so when out in the woods, people started wearing face masks on the back of their necks. The numbers of people killed have dropped considerably since the introduction of these highly-charged 'woodcutters' and with people showing eyes at the backs of their heads.

Despite these apparently effective protective measures, there is no doubt that in some places tigers were, and still are, serious predators of people, on a regular basis. Of course, the percentage mortality caused by tigers amongst people is very low, because there are relatively few tigers about. But each individual tiger poses a risk, and this is an issue that has to be faced if we want to keep these awesome but wonderful creatures in their natural habitat.

LEOPARD

The same Jim Corbett of tiger fame also describes the depredations of two maneating leopards in India earlier in the twentieth century, both males. Leopards are beautiful, nocturnal cats, much smaller than tigers (weighing 30–70 kg), and often about the same size as a person. Then, as now, leopards were much more abundant than their striped relatives, but leopard predation on people has always been much less common than that by tigers. Nevertheless, the numbers of people killed per leopard were still horribly impressive, and the two maneaters that Corbett recorded were alleged to have killed 400 and 125 people.

Leopard attacks were different from those by tigers. The victims were always taken at night, either inside their houses in the village, or on their doorsteps. In 18 cases described in detail, almost all victims were adults: ten women and eight men. Typically, a family would be asleep in their wooden house, and the leopard would break through the door or wall. It would kill with a throat or neck bite, and then carry the victim off to eat him or her several hundred metres away in the bush.

Such attacks still happen in rural India, and frequently. Here is a quote from *Asiaweek*, 24 April 1998:

> With a baby in her arms and eight-year old daughter Ritu trailing behind, Bideshwari Devi was making her way home when the leopard came out of the dark. It knocked the little girl to the ground, seized her leg and, crouched and snarling, began to pull her away. Bideshwari grabbed Ritu's arm, but tripped and fell down an embankment into a terraced field. Slowly her grip loosened. Weeping and shrieking in horror, the woman heard her daughter's cries as she was carried off into the night. Brought out by the commotion, Bideshwari's family set off in pursuit. Lighting their way with torches they followed spots of drying blood until they discovered Ritu's half-eaten body under a bush a hundred meters above the village.

This happened in Auri district in Garwhal, a region of northern India.

Later in the same village a 24-year-old woman, Sundari Rawat, who had stepped outside her home to relieve herself, was mauled to death, and then the wife of Balwant Singh Rawat, a neighbour, failed to return home. Balwant found her sandal on the road which then led him to her half-devoured body. The article mentions that 17 people were killed in Garwhal region in 1996, and 19 in 1997, all women and children apart from one drunk who was asleep on the road. The locals state categorically that attacks are always from behind, and (in contrast to the tactics of a tiger) always directed at the smallest and weakest of the party.

Leopard attacks occur not only in Asia. There are a few records of leopards killing people in South Africa, demonstrating that they are potential maneaters throughout their whole geographical range, although perhaps more in Asia than in Africa. But leopards are fairly common over a huge part of the world, and the actual risk posed by any individual animal is extremely small.

LION

The African lion is of similar dimensions to the tiger, and it habitually lives off a range of prey sizes into which a person would easily fit. As it occurs in countries where people often sleep with only little or no protection at night, it is not surprising that occasionally lions turn on us. A classic account is to be found in Patterson's *Man-eaters of Tsavo* (Patterson 1907).

Colonel Patterson was employed in Kenya in the early days of the colony, some time in the early twentieth century. His job was to provide security for the construction workers on the railway in the woodlands between Mombasa and Nairobi, and the impressive report showed that this was a massive undertaking. Over a period of many months the mostly Indian work force was terrorized by two male lions. Life was cheap in Africa, and the number of people taken was never recorded properly. As Patterson wrote '...they devoured between them no less than twenty-eight Indian coolies, in addition to scores of unfortunate African natives of whom no official record was kept'. Of one episode he writes:

> Hurrying to the place in daylight I found that one of the lions had jumped over the newly erected fence and had carried off the hospital water-carrier.... He had been sleeping on the floor, with his head toward the centre of the tent and his feet nearly touching the side. The lion managed to get its head below the canvas, seized him by the foot and pulled him out. In desperation the unfortunate water-carrier clutched hold of a heavy box in a vain attempt to prevent himself being dragged off... then caught hold of a tent rope and clung tightly to it until it broke. As soon as the lion managed to get him clear of the tent, he sprang at his throat and after a few vicious shakes the poor man's agonising cries were silenced for ever.... Dr Brock and I were easily able to follow his track, and soon found the remains about four hundred yards away in the bush. There was the usual horrible sight. Very little was left of the poor water-carrier – only the skull, the jaws, a few of the larger bones and a portion of the palm with one or two fingers attached. On one of these was a silver ring, and this, with the teeth (a relic much prized by certain castes), was sent to the man's widow in India.

Both lions were eventually shot. There is an extensive compilation of lion incidents in Africa in Charles Guggisberg's book on the species (Guggisberg 1962), with numerous cases of individual lions killing scores of people. In 1925, one animal killed 84 people in Uganda, another killed 44. In 1924, in Tanzania, 23 people were killed by two lions, and in 1908 more than 20 Africans were killed in one area in Mozambique. In 1950, in what was then Nyasaland (now Malawi), 14 people were killed. The tally is almost endless, but often with little detail: clearly many hundreds of people died this way, but most were unrecorded, they died as 'natives'. The large majority of victims were killed indoors, at night, with lions sometimes breaking through the walls of mud huts. Guggisberg mentions that, as a rule, lions grab a sleeping person by the head, and he or she is killed instantly.

Lion

In more recent years the threat of lion predation on people has clearly abated, perhaps because lions have been removed from populated areas. Also, just like tigers, many maneating lions have been selectively shot over this last century, and it could well be that this has caused the trend to diminish. It is possible that cultural transmission, i.e. one animal learning from another, plays a role in prey selection by these animals, as it does in many other species of mammals and birds. Perhaps lions have now learned from their elders that people do not fit into their normal spectrum of prey species.

However, the problem has not wholly gone away. In the Serengeti (Tanzania) in 1963, a tourist was dragged from his tent by his head, by a lioness. The man died soon afterwards. A similar incident was reported in newspapers in 1999 from Matsudona in Zimbabwe, when a young man was taken from his tent and partly eaten by a group of lions. In both cases the lions involved were immediately shot by the authorities. In the early 1990s there were many reports from the Kruger National Park in South Africa of lions killing people (Dr M. Mills, pers. comm.). These victims were taken from thousands of refugees from Mozambique who crossed the park, which stretches along the border between South Africa and Mozambique. They were people walking at night to escape detection by the authorities, only to suffer the attention of the many lions. Again it is impossible to put a figure to the numbers who perished this way – the parks authorities report that there must have been many.

Another beautiful cat, subject to many a story, is the sleek, reddish brown cougar, which stalks the woods and mountains of the Americas. Also called the puma or mountain lion, the cougar is large, not much smaller than a leopard, but there are nowhere near as many records of attacks on people by the animal. This may be because there are far fewer humans living in cougar habitat, and because these people live in conditions that do not expose them as much to cat predation as people in Asia or Africa. Nevertheless, some attacks on people do occur, and what is most worrying is that, in contrast to the decreasing predation by the other cats, cougar attacks are becoming more frequent.

Paul Beier wrote an authoritative analysis and summary of cougar problems (Beier 1991). He recorded ten people killed in 53 incidents between 1890 and 1990 in Canada and the USA. Interestingly, more than a third of these incidents occurred in just one area, on Vancouver Island, off the coast of British Columbia. Almost two-thirds of attacks were on young children, between 5 and 9 years old, and all people killed by cougars were under 13 years old, except for one adult who caught rabies after being mauled by a cougar. Of the ten people killed, eight were male. Over 40% of predation attempts on people occurred in the summer, presumably because that is when most people are out in the wilds: the large majority of incidents were away from houses.

Usually, cougars attack from behind, and, as with the tigers of the Sundarbans, people are often quite unaware that anything is amiss until they are hit. But in some cases the victim did see the predator coming, and almost every time that the child or person ran, the cougar pursued it and captured the quarry, so perhaps standing one's ground and fighting is a better strategy. One 13-year-old boy ran for 100 m before being overtaken and killed, but another 16-year-old boy ran, and he was being overtaken by the cat when he lost his boot and the predator stopped and ate it. The cougar was shot an hour later, its stomach full of old boot. In another case a very plucky woman screamed at two cougars coming for her, but they continued to approach and so she climbed up a tree. Both cougars climbed after her and one clawed her leg, but she managed to hit and kick them enough to make the animals abandon their course.

Beier's data are quite exhaustive, and he demonstrates that there were more fatal attacks during the last 20 years (six) than during the previous 80 years (four), perhaps because of increasing populations of both predators and prey. Yet cougar attacks are rare, and ten people killed over 100 years over the whole of North America is not that many, tragic as each individual case is. But attacks receive a great deal of publicity, and a cougar on the rampage gets many more column inches than, for instance, kills by domestic dogs, which are much more common. It is quite unclear why the mountain lions should be preying on people more often on Vancouver Island than elsewhere. Perhaps there is a 'cultural effect' amongst the predators – a social learning process.

SPOTTED HYAENA

The spotted hyaena is, despite its reputation, a large, wolf-like predator, often hunting the African plains and even the forests in packs. Even the admirers of hyaenas (myself amongst them) have to admit that they have a considerable crime record. It is now well established that hyaenas are not just the 'cowardly' scavengers of popular fiction: as we have seen in the previous chapters, they are very efficient hunters of large animals, and it should be no surprise that human beings are well within their spectrum of prey species. Nevertheless, scavenging is also important in their lives, and when I found human hair in faeces of hyaenas in the Serengeti, I was fairly confident that it did not come from a kill. More likely, hyaenas had eaten the dead bodies of Masai people placed out in the bush as tradition demanded.

However, hyaenas are also killers of people. They are larger than they first appear, and the ones that are involved in predation on people are especially heavy. Mr Balestra, a game warden writing in *African Wildlife* in 1962 (Balestra 1962), killed two hyaenas after incidents in Mlanje in Malawi. They weighed in at 72 and 77 kg, about the same weight or more than that of their victims. There was a human population of about 10 000 in that area of about 2000 km^2, and they were terrorized by hyaenas in the late 1950s. These predators killed and ate 27 people over 5 years. Many, but by no means all, of the victims were children. The timing was the same every year. When, during the hot dry season, people were sleeping outside their houses on their verandahs, the hyaenas walked up quietly, grabbed their victims, usually by the head, and dragged them off into the bush. There were probably only very few hyaenas involved, but they were sufficient to create a state of panic in the area until they were shot by Balestra.

Another incident from Malawi was reported in a newspaper clipping from *c.* 1972:

> HUNGRY HYAENAS GRAB CYCLIST. Three hungry hyaenas chased schoolmaster Nyirendas Luggage, as he was cycling to work, pulled him from his bicycle and badly mauled him before villagers answered his cries for help. The Malawi Newsagency reported the incident that happened in the Nkata Bay District on Lake Malawi. The hyaenas were tracked down by a game ranger who shot two of them dead and wounded the third.

Such events have also been recorded elsewhere in Africa. For example, in Kenya, the newspaper *The Nation* reported in February 1975:

> GIRL EATEN ALIVE BY HYAENA. A girl who fell asleep while looking after camels woke up as a hungry hyaena started tearing flesh from her face. She screamed but the hyaena continued chewing her left jaw. Reira Abdillah was saved by fellow camel tenders who heard her screams. She was admitted to Wajir District Hospital where her condition was reported as 'not serious'.

When I lived in the Serengeti in the 1960s, one of my colleagues, the well-known ecologist Tony Sinclair, had to abandon his vehicle after being stuck in a river. He walked home across the open grassland plains, some 20 km. This was in broad daylight, but he was followed by a pack of spotted hyaenas which attacked, and he survived by getting onto some rocks, out of reach of the hyaenas. Over the years several other scientists, too, were forced up trees by my spotted friends, and there was little doubt that those hyaenas were not just curious. Most of the intended victims felt that I had some responsibility!

A few years ago I had a first-hand account of an attack in a letter from an intended hyaena victim. Stephanie Simborg told me that in 1995, in a party of American college girls, she camped on the edge of the Serengeti in the Loita Hills, with several girls per tent. One night a spotted hyaena bit through the fabric, dragged her out into the bush, and only determined action from a Masai with a spear saved her life. She escaped with severe injuries to her face and arm.

BEARS

The huge, strong bears have always shared their northern habitats with mankind, and, awe-inspiring as they are, they feature as the classically frightening creatures of our wilderness. As we have seen, most bears are to a large extent vegetarian, even fish-eating carnivores, and one would not expect mankind to be part of their normal range of prey species. Nevertheless, attacks by bears have been reported, especially from North America, and they are evocatively summarized by Stephen Herrero in his book *Bear Attacks* (Herrero 1985). Both brown (grizzly) and black bears are involved. We also know that polar bears may attack and prey on people in the far north (in Canada six people were killed by polar bears between 1965 and 1985 (Fleck & Herrero 1989)), but there is much more information on the two other species. Brown bears cause many more problems than black, and Herrero reports 23 people killed by black bears for the period 1900–1980 in North America, as well as 'about twice as many' by grizzlies.

Grizzlies killed at least ten people in national parks, where animals are used to humans and the bounty they provide. In contrast, of the 20 deaths caused by black bears that Herrero studied in detail, only one was in a national park. In 18 of these 20 deaths by black bears, predation appeared to be the motive, and the animals were killing to satisfy hunger. The time of attack was known in 15 cases, and all bar one were in the daytime. Half of the victims were less than 18 years old, the smallest a 3-year-old girl taken near the door of the cabin, in front of her mother.

Geologists face occupational hazards in wild terrain, and they have been attacked by bears on several occasions. The victims were obviously adults, killed and partly eaten, sometimes even after being plucked out of trees. However, attacking bears are not always success-ful. Survivors have demonstrated that the animals can be fought off, especially black bears, which are the smaller species: a large black male weighs about 140 kg, while a large grizzly weighs more than 350 kg.

The following is one of the cases described by Herrero. One August Sunday in 1980 a party went fishing in the Glacier Park, Canada – two men called Ernest Cohoe and Bob Muskett, and three boys.

> At about 2:15 p.m. the men heard a crashing in the brush and then saw a brown bear very close and charging at them. The men shouted to the boys to run and then ran themselves. In their excitement each of the men ran in a different direction. The bear ran after Muskett, who only got about twenty feet before he fell. The bear stopped, stood on his hindlegs, looked

at Muskett but did not injure him. The bear quickly dropped onto four legs and ran after and attacked Cohoe, who was still running. Within seconds the huge bear clasped its powerful jaws around Cohoe's face and bit. Cohoe screamed and within a minute or two the bear attacked twice more. About a week later Cohoe died from his injuries.

Another case described by Herrero involved two hikers, again in a national park.

> They had just rounded a bend in the trail when they saw a grizzly bear, which was 'huffing and puffing', charging towards them from less than fifty feet away. Andrew Stepniewski had only a second or two to notice that it was a grizzly, but not a big one, when the bear grabbed hold of him. He screamed, yelled, and resisted for a few seconds, and then realizing that resistance was futile, he relaxed and put his hands behind his head. He thought that the attack on him lasted for only fifteen or twenty seconds and then the bear attacked Barbara Chapman. This attack lasted only a few seconds. Andrew remembers Barbara kicking at the bear and briefly trying to resist, and then the attack was over. Barbara was dead. Andrew was critically injured with head, facial, neck and body wounds. Despite his injuries, he managed to hike out to the busy Trans-Canada Highway in an hour and a half. Back at the site of the attack, the bear dragged Barbara's body about two hundred feet down a steep bank into heavy alder undergrowth and began to eat it.

There was a case in Algonquin Park in Canada, just a few weeks before I myself went hiking there, in May 1978. Four boys had been out fishing, and three of them were killed by a black bear, at about 5 p.m. One of them had run off to raise the alarm, and the others had been caught one after the other, independently, and shaken to death. Two of them were partly eaten, and their bodies were found next day. They had been cached in the undergrowth, as a classic example of surplus killing. The bear was shot; it was a large male, weighing 125 kg.

It appears, therefore, that the black bear actually preys and feeds on people, albeit comparatively rarely. But for the larger brown (grizzly) bear the story is different. In the vast majority of the reported attacks the animal was a female, going for people who came too close to her cubs. Sometimes, as in one of the cases mentioned above, the attack was then followed by the bear eating the victim. Before 1970, almost half the cases in the USA were from Yellowstone Park alone, where brown bears habitually fed in garbage pits. The bears there were also regularly fed by tourists, and several animals had lost all fear of people.

Black bear and cubs

There are some observations with clear details of bear hunting behaviour directed at people. These usually, but not always, involved the smaller black bear. An Indian guide, Harvey Cardinal, was walking in the backcountry of British Columbia, Canada, in January 1970, and tracks in the snow showed what happened to him.

> Cardinal was walking through the woods and had just passed a head-high mossy hummock when the grizzly attacked. The grizzly had been lying behind the mossy hummock and must have heard him coming. When he was just six feet past it, the bear circled behind the hummock into his tracks and hit him from behind. It ambushed him without warning. The attack appeared to have been sudden and deadly. The safety catch was still on Cardinal's rifle and his gloves were on. When found, Harvey Cardinal was frozen stiff. Most of his abdomen was gone. (Herrero 1985)

In the past the North-American Indians, on the whole, left bears well alone. But more recent tourists have been less cautious. For example, Herrero records 1028 black bear versus people incidents for the period 1964–76 from just one area, the Great Smoky Mountains National Park, in which as many as 107 people were injured. Thirty-two of them were feeding bears at the time. These incidents occurred in the tourist concentration areas, whilst in the same period there were only seven cases of injuries caused by black bears away from the roads, in backcountry. Obviously, therefore, many people are asking for trouble, but there is no doubt that in general both brown and black bears may constitute a serious danger to people.

Interestingly, brown bears are rarely a problem in Europe and Asia, although they share many areas with people. Bears are found fairly commonly throughout Scandinavia (and in eastern Europe), but only one person was killed there this century (excluding a few hunters who were mauled after shooting a bear). The further east one goes through Eurasia into Russia and Siberia, the more common bear incidents are, despite there being fewer people (Swenson *et al.* 1996).

WOLF

Grey wolves are carnivores that often occur close to man. Sometimes living in impressive packs, with their imposing stature and howls that can dominate an entire landscape, they would have a frightening reputation even without their criminal record.

Early records from North America gave rather damning evidence against wolves, and great pioneers such as Lewis and Clark had some hairy tales to tell (Lewis 1997). But later, scientists denied it all. David Mech from Minnesota is undoubtedly the world's wolf expert. He did his PhD on predation by wolves on Isle Royal, Michigan, and he has made a long, very productive career out of studying the animals in America, writing several books in the process. Mech states categorically that there is no evidence of any wild wolf in North America ever having attacked people 'deliberately', except when rabid. He mentions several newspaper reports, but, in all cases when these were followed up, such reports appeared to be false (Mech 1970).

Mech concludes that one cannot say that wolves are totally harmless to man, and perhaps an odd case has occurred of an attack on people. But if so, it would be a great rarity. In fact, just such a rarity occurred in 2001, when a wolf carrying a radio-collar attacked a child in Alaska. Fortunately, there were no lethal consequences, but it showed that one can never say never.

The almost complete absence of wolf attacks on people in North America is confirmed by several other authorities, and it must be genuine, not just resting on a lack of information. It is in striking contrast to recent history in Europe, where stood the cradle of Little Red Riding Hood. That fairy tale is based on actual horrendous incidents, which were not that rare either. Why wolves in Europe (and Asia as well) should behave so differently from those in North America is still quite unknown – but the data show indisputably that wolves were (and still are) regular predators on humans, often on children.

In early 1996 I was working in Belarus, participating in a radio-tracking study of European mink. We lived in Zadrach, a small village close to the border with Russia, near Gorodok. It is very remote, with about a dozen families of peasant farmers; there were no cars (just a couple of tractors), no telephone and no shops. People walked or went on skis to go anywhere, and in winter, with over half a metre of snow, life was very difficult. Wolves were and are common, and they often raided the village in search of domestic dogs or livestock. On 21 February in the late afternoon, just 3 days before I arrived, 60-year-old

Michael Amosov returned from Zadrach to his house in the hamlet of Bolonitza. He was a man known to my colleague there, Dr Vadim Sidorovich.

Amosov had to walk for about 3 km through the forest, along a clear cart road. At least, he set off from Zadrach, but he did not arrive at his house. The next day, many wolf tracks were found at a site about halfway to Bolonitza, the snow was churned up and there was blood. However, the weather was very bad, and when I left 2 weeks later, Amosov's remains had not yet been found. There was no doubt in anyone's mind about what had occurred, because wolf predation just happens there; it is a fact of life.

Sidorovich told me that 2 months earlier, in December 1995, a previous wolf victim in the area had been taken in Hvoschno, about 15 km from Zadrach. A woodcutter of some 55 years old was out in the forest on his own, and when he did not return parties went out to search for him. Two days later the few bits that remained of the man were found, surrounded by wolf tracks – another victim. But perhaps the most harrowing incident took place only two weeks before that, when a 9-year-old schoolgirl was taken by wolves in nearby Usviatyda. In that case a teacher had kept her late at school, and she walked back home in the dark along a lonely track. Her father was worried about her being late, and went out to investigate in the dark. He found her head, the snow spattered with blood and covered with wolf tracks. Later, in his wild grieving fury, he shot the teacher.

These events happened recently near a village and in an area that I happened to visit. No one there collects the statistics, and the authorities have other things to do. But I could not help wondering how much more of this would be going on there in the endless wilds of Belarus and Russia, never reported except in the odd newspaper article. Sidorovich, a scientist with vast experience in the area, informed me that wolf attacks are not at all uncommon. There are many wolves, and people are surprised that anyone in the west should doubt that wolves kill people.

Such horrific events must have taken place in Europe for as long as man and wolves have lived there. Almost unbelievable to me now, they happened and were well documented in my own country, Holland (Geraerdts 1981; Poortvliet 1994). The date was 13 August 1810, near the village of Helden, only a few miles from where I grew up over a century later. Bartholomé Dahmen, 9 years old, was helping his elder brother and sister with herding a cow and a goat, about 100 yards from their home. It was eleven in the morning, close to the woods – and the three children had little warning when a large wolf ran at them from the trees. Bartholomé was attacked and he was dragged off into the wood. When his father was alerted he ran, desperately following the tracks across a brook. There he found the remains of his son, still warm. Immediately the mayor rang the church bells, and people gathered and followed the wolf into the wilderness, armed with pitchforks. Their effort did not bear fruit.

That same month (on Monday 27 August) also in Helden, two young sisters Maria and Judith Geraerdts, 10 and 4 years old, were helping on their parents' farm, pushing a wheelbarrow of turnips along the Land Straat at about eight in the evening. A wolf ran out at Judith, and dragged her off while the eldest child, Maria, could only run and save herself. Villagers spent the night searching with lanterns, but to no avail. Next morning someone found parts of little Judith's body. People were still talking about the incident when, on 9 September and in the same village, 17-year-old Jan Joosten just managed to escape a wolf by running inside his parents' house.

These incidents occurred in what is now Western Europe's suburbia, where there are no wild wolves within hundreds of miles. But such cases do not stand alone: they are part of a long litany of sorrow for those years in the south-east of Holland. The events are documented in the local archives and departmental dossiers, and in extensive notes by several mayors, researched in detail by the historian Gerrit Geraerdts. A total of 12 children were killed there just in 1810–11, and several more were injured but managed to escape. The age of the dead Dutch children ranged from 3 to 10 years old, and of five people who were attacked but escaped, two were children of eight, the others were teenagers (aged 15 and 17) and one was an adult male. Quite likely several more people suffered, but their fate did not make it into the annals of history. The cases in Holland had in common that the fatalities were all children, whilst older and stronger people were attacked but escaped. All incidents happened in daylight, and mostly in summer.

During a recent stay in Estonia, I was fortunate to make contact with Ilmar Rootsi, an amateur historian with a keen interest in wolves. He had studied the archives of the rural Lutheran parishes of Estonia for the records of deaths that occurred there, also covering the nineteenth century. The causes of death are carefully registered when known, and Rootsi wrote down the frequencies with which wolves were mentioned. Despite the fact that this was some thousand miles away from the Dutch events, there are some remarkable similarities. His accounts (Rootsi 1995) provided cold data on predation in an almost randomly chosen country in Eastern Europe, and they give the scale of suffering, documenting the deaths of many children and even adults.

Only a few hundred thousand people then lived in the small northern country of Estonia, most of them in towns and far fewer in the countryside. But in the records of the nineteenth century Lutherans, Rootsi found that between 1804 and 1853 as many as 111 people had been killed by wolves, all away from the towns and villages. Almost all (108) of them were children below the age of 17. The average age of the wolf victims was just over seven years, with slightly more boys than girls killed (59 versus 47). About three-quarters of the incidents took place in the district of Tartuma, in north-east Estonia, near Lake Peipsi. The total number of registered wolf deaths in eighteenth and nineteenth century Estonia was 136. Casualties occurred very patchily, and there were clear outbreaks of wolf attacks. For instance, there were major waves of wolf predation in 1809–1810, and in 1846. In one parish alone, 48 children were killed between 1808 and 1853, with 36 killed in 1809.

Rootsi showed a clear seasonality in wolf predation on people, and the following chart gives the casualties by month:

J	F	M	A	M	J	J	A	S	O	N	D
5	14	6	10	15	14	28	23	9	2	3	7

These figures indicate a striking peak in late summer, just as for the wolf predation in Holland. Rootsi gives two reasons for this peak. Firstly, children are outside in summer, playing or helping their parents on the land, and secondly and most importantly, it is the time of year when the she-wolf has to provide for cubs, the time of highest energy requirements. This seasonality in predation contrasts sharply with that of attacks on people by rabid wolves: in nineteenth century Estonia there were 82 of those, with 37% in spring, 24% in summer, 4% in autumn and 35% in winter.

The data also showed that at least some of the wolves definitely selected children as prey, at the same time ignoring the cattle that the children were herding. Importantly, the attacks were usually carried out by solitary wolves, not packs, with clear evidence of idiosyncratic traits (just one individual specialist marauder concentrating on one kind of prey). For instance, there were two cases where three and six children were taken over a relatively short period, and in each case when people did finally manage to destroy one female wolf, there were no more victims. As in Holland, people were confident that they were dealing with healthy, wild wolves, not with domestic dogs or rabid animals.

Europe is large, and in the past wolves occurred almost everywhere; my cases from Holland and Estonia are only examples, with the stories repeatable elsewhere and many times over again. Similarly nowadays, my recent Belarusian examples are likely to represent victims in vast areas of Eastern Europe and Siberia. There is little doubt that the role of wolves in fairy tales and horror stories is well founded in fact.

Wolves also take their toll in Asia. There are records of hundreds of victims of wolves in Kipling's days in India (nineteenth century), and predation on people is still common now. In November 1996 a paper on Indian wolves was presented to a conference in Japan by Dr H. Satish, fresh from the field, with well-documented evidence (Satish 1996). In the State of Uttar Pradesh, 22 people had been killed by wolves that year, most of them children (78% of all victims were between one and four years old). There had been 74 recorded attacks. The government paid compensation to relatives, the equivalent of about 300 US dollars for each victim, and there were some attempts at wolf control, resulting in 10–20 wolves killed per year. Remarkably, the wolves did not kill many livestock; the explanation offered was that all livestock, even sheep and goats, were continuously guarded, whereas children often walked about alone.

Clearly, there are many areas in North America and also in Europe where wolves are no danger to man, where an encounter with wolves is a thrill but not a dangerous one. But the records show that the spine-chilling effect that howling wolves have on us is not without foundation.

The wolf's close relative, the coyote, also has a reputation in North America, and there is good evidence that they occasionally attack people, especially in suburbia. A recent paper documented 34 cases, in which 27 adults and 14 children (almost all under-fives) were injured (Baker & Timm 1998). Fortunately, those incidents rarely had fatal consequences, although earlier one 3-year-old girl was killed by coyotes (Carbyn 1989). Mostly, these attacks occur in areas where coyotes have lost their fear of people, especially when they have been fed deliberately, nearby houses.

DOMESTIC AND CAPTIVE ANIMALS

Attacks on people by domestic animals are somewhat marginal to the theme of this chapter, but they deserve a brief mention. In North America and Britain not a year goes by without press references to domestic dogs injuring or killing people, especially children. In Britain there is now a law to protect the public from dogs, the Dangerous Dogs Act 1991, with detailed obligations for dog owners. In the US dogs killed between 18 and 20 people in the 1980s, and they inflicted substantial injuries on about 200 000 people per year (Sacks *et al.* 1989). Some breeds are much more prone to aggressive behaviour than others, and not just the larger ones such as Rottweilers or German Shepherds. Many accidents occur with smaller dogs such as pitbull terriers. Several of the problem breeds were originally bred for aggression, which is the reverse of the aims of most other processes of domestication.

Similar problems between carnivores and people also occur when wild carnivores are kept in captivity. Keepers have been killed by tigers and lions, and even small 'tame' Eurasian badgers have attacked and seriously injured people (including myself). All such occurrences have been unwittingly engineered by people themselves in one way or another, and they are not likely to affect the relationship between man and wild beast. The aggression directed at mankind here is probably a social behaviour pattern, rather than predatory; the pitbull terrier or pet badger treats people as it treats its own kind, without any of the usual fears of man.

PREDATION ON PEOPLE: A SYNTHESIS

Anyone who has been struck by the beauty, power and fascination of carnivores would like to wish away something as abhorrent as maneating, or think of it as an occasional aberration at worst. However, maneating is on reliable record for a wide spectrum of carnivores, of different felids, canids, hyaenas and bears. The predator of man has many faces, even though these faces have much in common.

Nevertheless, although maneating is widely distributed both taxonomically and geographically, it still is a relatively rare occurrence, which is growing rarer still. In fact, it is astonishing that it does not occur more often, given the vast numbers of people in the areas where predators live, given the size of people compared with that of other prey, and given the overall vulnerability of humans when out in the bush.

In general, maneating is only an infinitesimally small factor of mortality amongst people. Other causes of death are incomparably more important, and deaths caused by carnivores are totally insignificant compared with those caused by diseases, slaughter on the roads or domestic accidents. The rarity of maneating is related to the rarity of the perpetrators. Yet individual large carnivores are a substantial risk, and it is that risk that is relevant to our behavioural response to the animals.

Maneating is not restricted to only the most carnivorous of carnivores. True, several members of the most carnivorous family, the felids, are known maneaters (tiger, leopard, lion, cougar, jaguar), but then most of the more omnivorous bears are also culprits. So the natural diet of a species is not a very relevant indicator of its danger to people. But one obvious characteristic stands out: size is probably the most important distinguishing character of maneating species, and the fact that four carnivore families are in no way involved (martens, raccoons, genets, mongooses) is probably a reflection of a lack of large species amongst them.

Size may be critical for maneaters, but rabies can render even the small carnivores dangerous. Maneaters as well as rabid carnivores occur on all continents except Australia (which is perhaps one factor that contributes to the more relaxed way of life down under).

Social organization also does not give much of a clue as to the danger that a species poses to us: some of the most murderous species are highly social (lion, wolf, spotted hyaena), whereas others are totally solitary (bears, tiger, leopard, cougar). But interestingly, in most cases involving the social species, they attack as solitary hunters, so maneating can be said to be a largely solitary crime.

There are some general, cross-species characteristics of predation on people. Most of the attacks are in daylight and outside (probably because of the activity patterns of people), although some predators (especially leopard, lion and spotted hyaena) may attack at night and inside houses. Children are much more vulnerable than adults, which may well have special relevance for our anti-predator behaviour, as I will discuss in later chapters. Most of the attacks on people outside are from behind – although this is not true for attacks by bears, which may often start as a defence of cubs.

There is a curious detail in the distribution of maneating behaviour: in many of the carnivore species concerned there appears to be a clear regionality of the habit. Wolves do not kill people in North

America but there are many cases of attacks in Eurasia, and bears in Europe appear to be quite harmless, while they are definitely not so in Canada and the USA. Cougars harass people especially on Vancouver Island, and tigers slaughter humans in the Sundarbans more than elsewhere.

This regionality of maneating suggests that either there are 'cultural effects' amongst the predators, with animals learning from each other, from parent to offspring as well as across lineages, or there may be a genetic element involved. There also appear to be definite outbreaks of maneating, with periods of many victims, which suggests that a 'cultural effect' is perhaps a more likely explanation than a genetic effect. The pattern of occurrence of such outbreaks makes it unlikely that the regional differences in maneating are caused by variations in the behaviour of people, nor do they appear to be caused by deficiencies in the usual prey base of these predators.

It is often assumed that maneaters start the habit when they are somehow disadvantaged, injured, or old and decrepit, then continue once they have discovered how easy and valuable a person is as prey. This may be the case at times, but it cannot explain every case: many of the predators concerned in recorded attacks were prime specimens when shot. It is likely that some carnivores chance upon people as prey, just because humans and their usual prey category have much in common, and they happen to find a person in a vulnerable situation; perhaps also, they were less afraid of people to start with. Once carnivores have experienced a person as prey they will learn fast, and other conspecifics will learn from them. Only a fast and radical management response will halt the process.

Despite all these gory details, and despite all the alarming publicity, we are only talking about a relatively very small number of incidents and people. In population terms, maneating is totally insignificant both to people and to the predator. Moreover, for several species of maneating carnivores the number of incidents is on the wane, probably because of better protection and prompt action by sharp-shooting hunters. Extrapolating this trend back into history, it seems likely that in the dawn of our evolution, when we were far less protected than now, predation must have been a much more important factor than it is now. This in turn means that predation is likely to have had a significant influence on the evolution of our behaviour and on our reactions to predators, including the disproportionate amount of publicity that we give them.

In future, though, we do not have as much reason for concern about the safety of our children as people had a few generations ago. Much of the element of danger from predation is being removed from our wilderness areas. Some of us may regret that, but there is no doubt that for the average villager in India or Africa it is a change for the better.

Wolves and sheep

5

Competitors and carriers
Predation on game and livestock

Conservationists will often deny that carnivores do damage to our interests. However, if friends are accused of committing some misdemeanour, it is obviously good policy to find out exactly what the crime was, before jumping to their defence. I am taking the same line on carnivores, and in this chapter I will detail more of their alleged wrongdoing. By doing this I am not taking an anti-carnivore stance, but I want to tally the counts against them, before putting it all in perspective in the following chapters.

Recently I was following the blood-stained track of an otter that had dragged one of my ducks through the snow and out of my garden in Scotland. I happened to be pleased about that event, because I thought it was very exciting to have an otter on my own doorstep, but not everybody would have been. On that occasion I was just one of the latest losers in a long, unending battle between carnivores and us. Before me, people have lost their livestock to the sharp-toothed forces of nature since time immemorial. Wild predators have cheated us out of our dues from hunting and farming ever since we took up the cudgel and chased our quarry in the wilds. Carnivores have been bothering our species in many ways, quite apart from eating us.

In this chapter I will discuss some of the evidence for damage to man's interests by carnivores, including damage to game, livestock and other resources, or to our health. It may be direct, by predation, or in a more roundabout way, such as through disease. An estimate of the cost of such injury to our interests is far from easy, because we have to bear in mind that we have already quite effectively protected ourselves against many potential infringements, often without even thinking about it or about the cost. Because of these defences we do not see so many actual misdeeds of predators any more, but we still have to carry the expense of protection. Put your hens behind chicken-wire to keep foxes away and you will never see any kills, yet you have to pay for the wire.

Even when we do feel bothered by wild animals, this does not always mean that our resources are actually being threatened by them. Perceived infringements may occur only in our minds, because we some-times do not see the whole picture. For instance, people often have the basic, simplistic idea that any game eaten by a wild animal cannot be taken by us ourselves – therefore there must be competition, as the predator is stealing our prey. This may be correct in some cases, but is not always.

It takes a long ecological explanation to show that competition is a much more complicated issue than it appears at first. Let me give an example. In many places numbers of deer are limited by a shortage of food, i.e. there cannot be more deer because of lack of fodder. In con-sequence, the removal of an individual by a wolf merely frees a living space for another deer, and that animal can then be taken by a human hunter. In such a case, wolf predation does not affect our hunting suc-cess, because if the wolf had not killed the deer it would have died of starvation. Also, if game is being taken from a population by predators, one may see an overall increase in reproduction, and thereby, the prey population compensates for losses. This is just one of the scenarios that occurs, and I mention it merely to show that the simultaneous use of resources by animals and ourselves does not automatically mean com-petition. In the end, however, whether the damage is real or imagined may be immaterial to people's attitude to carnivores: they are seen as a threat.

The assessment of damage to livestock is usually more straight-forward than that of competition for wild game, and whatever a preda-tor takes from our pastures comes off our plates. But, as I will show below, even in this case there are times when the infringement of our interests is apparent rather than real.

IN PURSUIT OF GAME

To human populations in days of yore, wild animals as game were an all-important resource. Now, however, the relevance of game to our survival has become more marginal, and it is usually an object of sport rather than food for the majority of human societies. But the question of competition between man and wild predators is still important, and the consequences for management are far reaching. What exactly is the evidence that carnivores rob us of the rewards of the hunt?

Examples come from many directions, from large game down to small birds. In Alaska, the ecologist Bill Gasaway and his colleagues addressed the long and acrimonious debate between hunters and conservationists over the effects of predation by wolves, and by brown (grizzly) and black bears on moose (Gasaway *et al.* 1992). Their critical study over 8 years shed some light on this, which was not of much comfort to the conservationists. It suggested that predation by these large carnivores was responsible for keeping moose populations at an average level of only 148 moose per 1000 km^2. When predators were shot and trapped, the same areas carried an average of 663 moose per 1000 km^2. In terms of moose harvest by hunters, the areas under predator control yielded 20–130 moose per 1000 km^2 per year, compared with 0–18 moose per 1000 km^2 per year when wolves and bears were left alone.

However, the Alaskan study conceded that the matter was complicated by the presence or absence of alternative prey (especially caribou). There is additional and illuminating information from elsewhere. In one area, the 500 km^2 Isle Royale in Lake Superior, Michigan, researchers have a fabulous opportunity to closely monitor wolf and moose populations, and this has now been going on for over 30 years. They found moose to be by far the most important prey for the wolves. Populations of both species are not managed in any way, and they show huge changes in numbers over time. When wolf numbers were at a low level, sustained over several years, probably as a result of disease, the moose population surged, then suffered an 80% crash due to starvation (Peterson 1999). The observations suggested that wolves have an important, regulating effect on the moose. Isle Royale may be an unusual situation, however, as wolf numbers seem to be highly susceptible to disease (perhaps because of some inbreeding effect in their isolation) and well below 'carrying capacity' in most years. But it seems highly probable that wolves can at times suppress moose numbers, and thereby not only compete with human hunters but also prevent moose populations from reaching catastrophic densities.

In Britain there is little question about the damage that we are suffering from carnivores, at least in the minds of those who manage populations of the Scottish red grouse. Grouse are game species valued especially because of the sport they provide. They are also excellent to eat, but grouse shooting is much more important as a sport than as a provider of protein. A major report on grouse management from the British Game Conservancy in 1992 concluded: 'Given a suitable habitat, legal predator control is essential to limit the effects of predation and to achieve a sufficient density of grouse for driven shooting. Only then will the economic, social and conservation benefits of grouse management be fully realized' (Hudson 1992). The predators in this case are the fox, crow and hen harrier.

Whether the claim is justified is not immediately obvious. It may well be, but the data and statements to back it up are not always convincing, and they may be open to other interpretations. For instance, in the report, data are mentioned on the year-to-year relationship between the number of foxes killed and the number of grouse shot, for six Scottish estates. On three of the six estates there was a negative correlation, showing that more foxes were shot at times when fewer grouse were present, and there was no obvious trend in the other three estates. The author drew the conclusion that this is 'consistent with the hypothesis that predation is an important factor influencing the abundance of grouse'. It was assumed, that many foxes shot meant that many foxes had been present. This is not necessarily true: it may also mean that there were more keepers, who would also be managing the grouse populations well in other ways. Furthermore, it could be inferred that the estates were selected to demonstrate the conclusions drawn. Even if all these six estates had shown the same trends, and even if the keepering effort had been constant, the conclusion should probably have been that a good fox year is a bad time for grouse. This could be for various reasons, not necessarily because foxes killed the grouse.

However, another recent and elegantly designed study from the Game Conservancy in England (Tapper et al. 1996) demonstrated clearly that predators can sometimes have a devastating impact on the numbers of game birds, in this case grey partridges. In the experiment, foxes, crows and magpies were killed in several areas, and in others they were left alone. Over a 6-year period experimental areas and controls were also reversed. The results demonstrated beyond doubt that in a fairly typical, agricultural area of England, with arable land and hedgerows, partridge populations were strongly affected by predation. Predator control in spring increased the numbers of partridges in the autumn by a factor of 3.5, and the numbers of partridges breeding the following year by a factor of 2.5. Without quibbling over how much each of the predator species contributed to this effect, there was no doubt that foxes were a major component in the equation.

The exact contribution of each species is difficult to quantify and generalize, because removal of one species of predator may be partially compensated by increased predation by others (Korpimaki & Norrdahl 1998).

Nevertheless, we are talking about real competition between man and carnivore here. The grey partridge used to be Europe's most numerous game bird before numbers crashed with the use of pesticides, and in the 1930s more than 20 million partridges were shot in Europe every year. Perhaps there were also fewer foxes then, but that is one of several imponderables. One of the problems with fox predation is that often it is not density dependent, i.e. foxes continue their pressure on a prey species even when it falls to very low numbers. In low-density prey populations, therefore, foxes can have a disastrous effect (Leckie et al. 1998).

A study of fox predation on several species of duck in central North America estimated that annually about 900 000 ducks are taken by foxes (Sargeant & Arnold. 1984). It is quite possible that this predation has an effect on the harvest of waterfowl, but only several years of properly conducted experiments would enable us to estimate the magnitude of this. A more interesting result, however, came from a study of the effects of foxes on duck nesting success, combined with observations on the complicating effects of coyotes (Sovada et al. 1995). The researchers showed that where there are coyotes, ducks do much better than where there are none, because the coyotes are not particularly interested in ducks and keep the fox population at bay, thus allowing the ducks to breed.

Sometimes a 'natural experiment' can give a good clue about what is taken away from us by a competing carnivore. In the 1970s, the sea otter came back to the shores of the North American west coast after an absence of many decades. The Pismo clam is amongst its favourite prey, as well as being the target of a popular form of sport fishing. In each of four well-documented cases on separate beaches, the clam almost totally disappeared about one year after the return of the sea otters, and the fishing dried up completely. Similarly, the fishing of red and Dungeness crabs (also popular prey of sea otters) saw a very sharp decline whenever the predators turned up, and there was little doubt of who came out on top in man's competition with the sea otter (Kvitek & Oliver 1988).

If, indeed, predators take the quarry in front of our noses, how much does it cost us? There are a few cases where we can get at least some idea of this. In Scotland the sport of grouse shooting alone contributed £21 million ($34 milion) to the gross domestic product in 1990, and it provided the equivalent of 2323 full time jobs, including some 730 gamekeepers, who each cost about £15 000 ($24 000) per year in salaries and expenses. To be sure, gamekeepers also do a great deal other than predator control, but trapping and shooting 'vermin' is one of their main activities, perhaps taking up half of their time. Vermin include many species other than carnivores, but foxes are the most important adversary of the keepers, and predators such as wildcats, stoats and mink are on their list as well. It seems a valid conclusion, therefore, that, whether predation by these animals on Scottish grouse stocks actually detracts from our grouse harvest or not, persecution of these vermin does cost a great deal of money. Taking keepers' salaries and costs into account, on the Scottish grouse estates alone, the annual price tag of carnivore persecution amounts to something in the order of £2–3 million ($3–5 million) per year at 1990 prices.

The picture I have painted of competition between carnivores and people over game is relatively simple, and perhaps deceptively so. There are important complications, and the role of carnivores in ecosystems will be discussed in more detail in Chapter 9. However, also at this stage it should be recognized that there are situations where carnivores have a distinct negative effect on game bags, although by no means always. We also find many cases such as the relationship between coyotes and mule deer in Colorado (Bartmann *et al.* 1992), where well-controlled experiments demonstrated that deer populations show density-dependent mortality. Whether the mortality agent was starvation or predation by coyotes (controlled by researchers), the outcome (in terms of numbers) was always the same, and coyote control had no effect on deer numbers. It is quite likely that such cases may even be in the majority, but this does not negate the others in which carnivores do influence hunting bag sizes. There seems to be little doubt that for the hunters amongst us, it is expensive to have carnivores around, however much we like to see them.

KILLING LIVESTOCK

Many of us will admire the sleek form of a red fox or coyote, when it trots through the fields in search of voles. But its sharp, wild teeth have also been an anathema to our sheep, goats, cattle and whatever else we nurture, ever since we started the animal husbandry enterprise between 50 and 90 centuries ago. Carnivores have been at it all this time, and when today I get on the Internet and search for predation on livestock

Common genet, killer of chickens

I get a massive response. We have had, and still have, a major problem, because a carnivore's interest in livestock is similar to our own.

I want to mention a select few recent studies from all over the world, on assaults by predators on the animals we keep and farm, involving livestock as diverse as sheep, cattle, camels and farmed fish. The studies may present exact data, but when contemplating these, we have to keep two points in mind. Firstly, the vast majority of cases of carnivores killing livestock never make it into the statistics and are never written up: in lots of places a few hens or sheep are killed, and even if the remains are found nobody makes a fuss about it (this includes losses of many of my own hens and ducks). Secondly, we have already protected ourselves with reasonable effect against the foe, with fences, shepherds, dogs, guns and whatever else. This protection means time and effort, and therefore cost, even if the predators are out of sight. In consequence, the cost of damage goes quite a bit further than what we can estimate on the spot.

During a project in the late 1970s in the desert in the north of Kenya, not too far from Lake Turkana, I crawled out of my tent in the early morning. I was camped under a large acacia tree next to a dry riverbed, miles away from any town or road. Birds were singing, and above that I could hear cattle in the far distance. It might be wild, wild country, but I was never far from livestock there. Soon I was off for my usual sunrise walk, to watch birds and mammals, and to look at tracks in the sand.

I turned away from where I heard the cattle and strolled slowly through the bushes, stopping to look through my binoculars, spotting some kites and a jackal. Twenty minutes later a vulture started to come down, just ahead of me, and immediately I had to be on the alert: where vultures land there may be killers, and in the night lions and hyaenas had been calling. I saw flapping wings, there were blood-stained tracks in the sand, and there was a smell of rumen contents. Under a bush were the remains of a calf, with just about half of it left.

It did not take me long to trace what happened. Two spotted hyaenas had eaten most of the carcass, and apart from jackals no other predator had been near. Patterns of subcutaneous bleeding suggested that the calf had been killed by the hyaenas, and not just scavenged. I was lucky to be early, because, with all the vultures, in another hour or so only the large bones would have been left, and tracking would have been very difficult.

Later that day I talked to some nearby herdsmen, people of the Gabra tribe whose huts were a few miles away. I counted their herds, somewhat surreptitiously, and they gave me some more figures. They told me about a donkey killed by lions, and two goats taken by hyaenas over the last 3 weeks. They gave me time and place, and I was able to check on leftover bones, on old tracks and scats. The people also told me that the previous night a calf had gone astray on the way home from grazing, but they did not know its fate. I did.

At the time I was spending 4 months in the desert country, amongst the pastoralist and camel-keeping tribespeople (Turkana, Rendille, Samburu and Gabra). I was charged with a survey for the United Nations Environment Programme (UNEP), to find out how much damage the nomadic tribes were suffering from wild predators. It was the old African problem again: the government and the tourist trade wanted a hands-off policy for wildlife, but the graziers protested strongly, claiming that their life was made impossible by lions, hyaenas, jackals, cheetahs and many others. Who was right?

I talked to the people, spending hours sitting under shady trees next to their 'manyattas' (hut circles), watching them guarding their herds of sheep, goats, camels, cattle, walking many miles every day to the waterholes, and to the few patches of edible vegetation. I estimated numbers of cattle, and I checked information from the elders and the herdsmen wherever I could. I counted lions and hyaenas (by playing hyaena feeding calls from loudspeakers at night), and analysed hundreds of scats. A bleak picture emerged, a story of incessant human failure in keeping predation at bay.

When I finally tallied all the data and made my estimates (Kruuk 1980), I reckoned that these pastoral tribes lost between 2% and 10% of their livestock per year, mostly to the lion, spotted hyaenas and black-backed jackals, but also to the cheetah, wild dogs and striped hyaenas. The percentage may not be high, but in those arid lands it was a small fortune. The losses occurred despite all the hard, day-long and

Black-backed jackal

everyday efforts of men, women and children in herding and fencing. Most animals were lost when grazing, in the daytime, but 10% of kills took place right inside the 'bomas' (stockades) at night.

For these nomads, one of the costs of predation on livestock was the expense of building bomas, work done mostly by women. Costs included not just the considerable number of working days, but also the thousands of thorn trees that had to be cut around the settlements, causing large areas to be denuded. Each household uses 70–100 trees per year for fencing, a large amount because of the frequent moves of bomas to wherever the grazing is.

This story of carnivore menace is quite typical for many areas I have visited in Africa. An example has been reported from Tanzania, where a survey in areas around national parks showed that about 10% of the people living there reported recent damage from wild carnivores to livestock and poultry, some of it quite substantial (Newmark *et al.* 1994). Carnivore predation on livestock is a major factor out there – and that comes on top of the threat from these animals to human life.

All this was in Africa, in wilderness where the reign of carnivores is legendary. But these things also happen in America. A 1976 federal bill allowed for an annual government compensation of up to $500 000 to be paid to Minnesota farmers who had lost livestock to the many wolves in that state (Macdonald & Boitani 1979). In fact, annual compensations for loss to wolves until 1989 averaged about $24 000, mostly for calves, sheep and turkeys (Fritts *et al.* 1992). In 2000 this figure had risen to $67 000 per year, and together with wolf control the government was expected to spend over $400 000 p.a. in the years following this, just in Minnesota alone (Mech *et al.* 2000). David Mech and his colleagues estimated that per year about 1% of Minnesota farmers within wolf range suffered from wolf depredations; there was no obvious characteristic of farms that were affected.

Another census was done in Iowa in the 1970s, covering 1251 sheep farms. It found that coyotes were responsible for 35% of all losses of lambs, and for 22% of all losses of adult sheep. In total they killed 3% of the state's sheep annually (Schaefer *et al.* 1981). In Kansas, however, in the same period, a large-scale study found that predators killed less than 1% of the sheep per year (Robel *et al.* 1981), considerably less than in Iowa, although one still talks in terms of many hundreds of animals.

In California, predation on sheep was studied at a university experimental farm. Three per cent of the lambs and 1% of the ewes were killed by predators each year. However, if the managers assumed that those animals that just disappeared had also been killed by predators, then the figures increased to 10% of the lambs and 4% of the ewes. In 90% of the cases coyotes were the culprits, with the rest of the animals being taken by dogs, bears, cougar and golden eagles (Scrivner *et al.* 1986). Excluding the uncertain disappearances (the 'black loss'), the value of stock lost to predation was estimated at $6200 per year for just this one farm.

There are similar examples from Europe, almost on our doorstep: take Italy, for instance. In the wilder hills of Tuscany, wolves are quite common. So are sheep, vast herds of them, often brought in summer from southern Italy to graze the lush Tuscany pastures. Dense flocks are closely guarded by strong, tough shepherds, the animals corralled in at night behind movable fences and nets, each flock attended by two or more Abbruzi sheepdogs. These dogs are a phenomenon in themselves. They are aggressive monsters almost the size of wolves, wearing collars with long spikes pointing outwards to protect them against the predators. At night the sheepdogs stay with the sheep and around the corral, and there are fires with people sleeping nearby. The barriers seem impregnable. Yet despite all that, wolves manage to take a formidable toll of the sheep in Italy, year after year – and not just sheep, but also cattle and horses.

Luigi Boitani, now professor at the University of Rome, has made the Italian wolves and their problems a lifelong interest. One early morning, when we walked through the hills together, we chanced on one of the sheep camps, where all the people and some 300 sheep were gearing up for the day's long move across the hills. Dogs barked and bared their formidable teeth at us, shepherds were surly. The reason was not difficult to guess, with parts of three steaming, mutilated sheep carcasses pushed against the fence, and clear trails of wool leading off into the bush. For Boitani it was a familiar sight, and from the way in which the sheep had been killed and dismembered there was little doubt that large, wolf-like canids were the culprits.

This was just one event, one statistic. Often, other mortality amongst livestock is more important than predation, e.g. calves wander off and animals starve, and the consequences of that are less spectacular. But that does not take away the fact that predators can be very damaging indeed. The Italian government, in an effort to maintain its wolves as well as its voters, pays compensation to shepherds who are losing sheep or other livestock to wild animals. Using official data, Luigi showed that from 1991 to 1995 the government paid an average equivalent of $327 000 per year in compensation for livestock predation (compensation is paid only after assessment of damage by a team of experts, when it can be demonstrated that wild animals are involved and not, for instance, feral dogs). Wolves killed 0.35% of the total regional livestock production (Paolo & Boitani 1996), including on average 2550 sheep per year, and it has been estimated that there are some 200 wild wolves in Italy. To put this in some perspective, these areas have a huge density of sheep, sometimes up to 77 per square kilometre. On average, wolves kill three sheep per attack, but there have been reports also of 'surplus killing' in Tuscany, in which in one night as many as 264 sheep were killed in one herd (Macdonald & Boitani 1979).

Another, recent study of compensation claims following wolf predation in Italy showed that in the central Italian region of L'Aquila more than 30% of 'management units' (farms etc.) suffered damage from predation (Cozza et al. 1996). Almost all attacks came from wolves and a few from bears. Some farms had an average of eight incidents per year. Half of the claims were for sheep, one third for horses. Clearly horses and cattle were more vulnerable than sheep, because sheep were herded and horses and cattle were just left alone, often even overnight. All this is pretty expensive for the Italian tax payer, as is frequently and firmly pointed out.

In Scandinavia most livestock problems are caused by bears, lynxes and wolves. Bears are making a comeback in Norway, and with some protection they now occur again in many areas where they were absent before. Unlike the Italians, Norwegian farmers have become used to letting their sheep graze unattended. Not surprisingly, the bears soon learned that there was food available, and problems arose. The Norwegian Institute for Nature Research (NINA) studied three flocks of sheep in a 100-km^2, hilly and forested area (Knarrum et al. 1996). In 1994, they put 577 radio collars on ewes, lambs and rams, almost half of the total sheep population in their study area. Each of these collars had a mortality switch, which meant that the signal changed when the animal died, and the researchers could go out and establish what had happened.

The results of this study were dramatic. After the sheep were released in spring and before they were gathered up again in autumn, bears had killed 22% of the ewes, 4% of the lambs and three of the four rams. In addition, a few lambs were killed by lynx. The rather morose conclusion of the scientists was that 'traditional Norwegian sheep farming in permanent bear-areas is so difficult that heavy political and management tools have to be taken into use to make the co-existence of bears and humans in these areas possible'.

Bear predation on livestock in Europe is not just confined to Scandinavia. The Italians also see a lot of it. Wolves do much more damage than bears there, yet in 8% of attacks on livestock in central Italy bears were the culprits. Bears focused their attention on sheep and goats in 68% of 389 recorded cases, the other attacks being on cattle, horses and donkeys. The predators also demolished several beehives in the process. Usually bears kill only one animal per attack, but sometimes they get a surplus-killing urge, and in Italy an individual bear may slaughter up to 36 sheep or goats in a night (Fico et al. 1993). In northern Spain bears can be just as much of a nuisance as in Italy, if not more so, because they take cattle and horses more often than sheep and goats (Clevenger et al. 1994).

Wolverines are not much larger than a badger, but they can be ferocious killers. They do not confine themselves to just wild rodents and ungulates, but also take livestock, especially sheep and reindeer.

In Norway wolverines had been almost exterminated, but in the 1990s the population built up again to about 150 animals (Landa *et al.* 1997). In one area alone, the Snøhetta Plateau in central Norway, livestock losses to wolverines increased annually up to 1994 to about 2500 ewes and lambs, out of a total flock of about 38 000 (Landa *et al.* 1999).

As wolverines are protected the Norwegian government pays compensation for damage caused by the animals. A lamb for instance, with a market value of $110 equivalent, attracts a compensation of $170 if it can be demonstrated that it was taken by a wolverine. The overall result is that now in the 1990s an annual amount of almost one million dollars is paid for wolverine damage to sheep farmers alone (Landa & Tommerås 1996). This does not allow for other damages and the costs of administration and verification, a hugely expensive exercise. Interestingly, despite the high level of compensation the farmers still loathe wolverines, and (illegally) kill them whenever the opportunity arises.

Another species that attracted protection from conservationists and repaid this in a rather unwelcome currency is the otter. In the 1960s and 1970s the species disappeared from many countries in Europe (because of pollution), and almost everywhere it was put on the protected list. Since the 1990s it has been making a big comeback, it features prominently on television and in books and nature magazines everywhere, and it is a conservation success story. Those magnificent, lithe creatures are back again in many of our waters.

However, foremost amongst those who are not cheering this development are the fish farmers in Central and Eastern Europe. There are hundreds of fish farms there: carp especially is a highly prized product, being the favourite Christmas fare in several countries. Otters take large numbers of fish, mostly carp, from these fish farms, in quantities that are not easy to assess. The best data come from Austria, where government compensation is paid for confirmed otter damage, and where a highly skilled team establishes the cause and the extent of damage. Annual amounts paid out to fish farmers started to become substantial in 1989. In 1995 they had reached the equivalent of almost $200 000 (Bodner 1998). The impact of otters in other, poorer countries, such as the Czech Republic, Slovakia and Hungary, is likely to be higher still.

Obviously, the list of examples of carnivore misdeeds towards our livestock enterprises could fill several books. They include otters killing several scores of hens, ducks and geese around the crofts on the Outer Hebrides island of North Uist (*Press & Journal* 23 October 1996). A headline in this newspaper later summed it up simply – 'Killer otters

Eurasian otters

should be shot'. Other examples from the 1980s and 1990s include snow leopards taking 2.6% of the livestock annually in a poor village area of Nepal, including goats, horses and yaks (Oli *et al.* 1994).

On my doorstep and throughout Scotland, thousands of lambs disappear every year, mostly unrecorded but accepted as black loss. An unknown proportion of them is killed by foxes, and whether justified or not, foxes get almost all the blame for the black loss. In consequence, considerable resources go into fox control. For instance, in one year (1987), which appeared to be quite typical, farmers and government paid out £75 000 ($120 000) to 'fox clubs' (Dr R. Hewson, pers. comm.). These are groups of enthusiasts who go out killing foxes, and it was assumed that this prevented a much higher value in damages to the sheep farming industry. Fox predation on sheep does not attract official government compensation to farmers, because foxes are not protected; they are pests and people are expected to make their own arrangements to control them. But arrangements such as fox clubs are subsidized.

There is no doubt, however, that much of this blame on foxes is unjustified. The large Scottish island of Mull has no foxes (and masses of sheep), and the black loss of lambs is very similar to that on the mainland (Dr R. Hewson, pers. comm.). It is likely that some of the ewes are 'bad mothers', and they will lose their lambs to starvation or chilling, if not to foxes. This is not to say that foxes cannot be major pests in sheep flocks, and particular individuals may specialize in taking many young lambs.

What may make a carnivore's inroads on our four-footed capital particularly galling is the indulgence of the predators in a behaviour that I discussed in Chapter 3 – surplus killing (Kruuk 1972b). We all know about it and I have discussed it in Chapter 3, the fox in the hen-house that slaughters everything and takes away perhaps just one of the victims, or the leopard or wolves that kill most in a herd of sheep in one single night. Noticing that a sheep or a hen has disappeared, and been eaten by the predator itself is one thing, but finding all one's efforts of animal husbandry put to absolute and total waste in one big orgy of slaughter is much more unacceptable.

Loss of livestock to carnivores is especially serious when it affects poorer people, as in the above case of pastoralist Africans who lost up to 10% of their livestock per year. Apart from the economic damage, there is also the inconvenience, and the emotionally upsetting experience of losing animals of which one has grown fond. The prevention of such damage is expensive, and if we were to calculate it we would find that it costs us fortunes in time spent guarding, in fencing and maintaining guard dogs. With this we are at the cutting edge of resource competition.

Whatever we decide to do about these fascinating carnivores that do so much damage, we have to face the extent of our losses as a fact. Then, grin and bear it is one possibility, compensation another, and carnivore control another.

TRANSMITTING DISEASE TO PEOPLE AND LIVESTOCK

One frightening scenario in which carnivores attack people is created by rabies. Attacks by rabid wild animals have nothing to do with predation, but they are probably much more frequent than predatory incidents, and they must have occurred since time immemorial. Rabies in wild animals often occurs in distinctive outbreaks, and sometimes a wave of the disease may travel a continent. The latest outbreak in Europe came from the East in the 1930s, and burned itself out in Western Europe in the 1980s (Macdonald 1980, 1995).

Rabies is a virus disease, attacking the central nervous system and especially the brain. The symptoms may be paralysis, followed by a quiet and miserable death, or alternatively, animals may be hit by the better known 'rage', which results in biting any other animal within range, including man, and thereby transmitting the disease. As far as we know all mammals of any size are susceptible, but some, and especially carnivores, get the disease much more often than others, and are much more involved in its transmission to people.

Foxes are the most common vectors of rabies in Europe and North America. But badgers also carry the disease, as do skunks, wolves, coyotes, raccoons, cougars and many other species, and they may in some areas be the dominant vectors. In South Africa the yellow mongoose is a major reservoir species, in Kenya it is the honey badger, and just about everywhere in Africa jackals are prone to rabies. Wild dogs also get it (rabies exterminated a population in the Serengeti (Burrows 1995)), and so do hyaenas (Mills 1990), as well as bat-eared foxes, arctic foxes and many others. If a rabid animal bites a person this means a high probability of death, unless the victim is treated with a vaccine before the symptoms manifest themselves – usually a time-span of 2–8 weeks.

A rabies vaccine was not available until Pasteur's spectacular work in 1885. Before then, rabies was an absolutely lethal factor in the carnivore/people mixture, and in many places it still is, albeit to a lesser extent. Very rough global estimates from the 1970s and 1980s are of an appalling 15 000 people killed each year by rabies, despite the existence of a vaccine. It is reported that as many as 10 000 people have died annually in India alone (Macdonald 1980). The global figure includes deaths from rabies transmitted by animals other than carnivores, such as vampire bats in Latin America. But such data serve to illustrate the magnitude of the problem. They mean that human deaths from rabies, acquired from carnivores, far outstrip the number of people killed through predation.

The overriding majority of actual transmissions to people come from domestic dogs (more than 90%), followed by domestic cats, and then wild carnivores, mostly foxes (Kaplan 1977). Domestic animals become infected after being bitten by wild carnivores or by other domestic animals. Several of the Estonian wolf attacks involved animals that were rabid, and a victim of a cougar in North America also died of rabies. But whether one blames the wild or the tame, it is carnivores that are the guilty party.

There are other pathogens fatal to people that are transmitted by wild carnivores, for instance the tapeworm *Echinococcus multilocularis*. This parasite lives in species such as the fox: for example in Germany some 6–11% of foxes may carry it (Berke & von Keyserlingk 2001), and in the Czech Republic over 60% (Martinek *et al.* 2001a). It is also found in wolves (Martinek *et al.* 2001b), raccoon dogs (Thiess *et al.* 2001) and many other carnivores and ungulates such as wild boar. The eggs appear in scats, and if they are taken in orally by humans the tapeworm may enter the body, and via cyst production give rise to the disease of alveolar echinococcosis. If the cyst happens to be in the brain, then the affliction may be terminal. Such incidents are severe, but mercifully not very common, despite the increased contact between people and animals, e.g. foxes living in towns (Eckert *et al.* 2000). Another species, *E. granulosis*, causis cystic echinococcosis, and is widespread worldwide, with frequent cases in Bulgaria, Kazakstan and China (Eckert *et al.* 2000).

In America raccoons carry a small roundworm (*Baylisascaris procyonis*), which may have very nasty effects in people as well as in other animals such as rodents. Contamination happens when people handle raccoon faeces or even when they handle other animals that have done so. The roundworm may cause severe encephalitis, which is sometimes fatal, especially in children (Boschetti & Kasznica 1995; Zagers & Boersema 1998).

Toxoplasmosis is caused by a protozoon, *Toxoplasma gondii*, and it appears that about one-third of all adults in America and Europe have been in contact with it, as they carry antibodies in their blood. It is a parasite that can have serious effects, especially in vulnerable people such as pregnant women, or HIV-positives (Jones *et al.* 2001), causing neurological symptoms, blindness or deafness, paralysis, convulsions and ultimately death. In this case it is especially cats that are the carriers, but it also occurs in mink and other carnivores.

The litany of misery about rabies, parasites, carnivores and man also extends to livestock. Rabies can have a significant effect on the health and survival of our farm animals, as well as on that of ourselves. In Latin America its impact is horrific, with $250 million per year damage to the cattle industry alone (Macdonald 1980). There, however, only a small part of the blame can be laid at the feet of the local carnivores, because vampire bats are the main carriers. At night massive numbers of them descend from their caves on sleeping animals, and often large numbers of these bats are infected.

In other countries carnivores are the most important vectors infecting livestock with rabies. Although the numbers affected are much smaller than in Latin America, there is, nevertheless, substantial damage. Between 1968 and 1973 in France, 697 cattle were diagnosed with the fatal disease. Rabies has been found in sheep, horses, pigs, and of course in domestic dogs and cats, but there are few exact figures (Kaplan 1977).

Talk about rabies, even about rabies in cattle, horses or dogs, evokes nightmare scenarios. For a farmer it is not just a matter of losing a number of animals, it is also the way they die, and the suffering involved that shapes our reaction. Take this rather clinical description of rabies in horses, from Colin Kaplan's book (Kaplan 1977):

> The symptoms vary from animal to animal. The minority develop the paralytic form of the disease, but most have an agonising death. As in other animals the first signs are changes in behaviour accompanied by a mild rise in temperature. The victim becomes anxious, and mental aberration gradually increases to a state of marked agitation. Sexual excitement may be intense. The upper lips are drawn back baring the teeth, and wrinkling the nose and lips. Rabid horses appear to be thirsty but cannot swallow. They shake their heads violently, foam at the mouth, grind their teeth and whinny frequently as if in great pain, lie down, stand up, sit like a dog, strain in futile attempts to pass dung, may lash out wildly with their hind legs, and show signs of severe colic. Although they will usually bite at anything such as a stick thrust at them, they do not show aggression, but may show a marked antipathy to dogs. Finally they go down and are unable to rise. At this time they may thrash about with their feet...

Fortunately, it now looks as if the battle against the disease can be won. At last rabies appears to be under control in Europe after an elaborate campaign to vaccinate foxes with an oral vaccine, distributed in the countryside in chicken heads. Previously, enormous amounts of money had been spent over many years in attempts to control rabies by killing the foxes, to no avail whatsoever. Oral vaccination appears to have solved the problem.

A quite different saga is that of another scourge of mankind, tuberculosis or TB. In livestock it is not in the same league of importance as rabies, because problems occur in a small area only, and the amount of suffering caused by TB in which carnivores are involved is far less than in the case of rabies. Nevertheless, it is a large and expensive issue.

There are different kinds of tuberculosis, but bovine TB is the one that concerns us. It causes consumption, the killer of hundreds of thousand of people throughout history. The perpetrator is the bacterium *Mycobacterium bovis*, thought to be under control with modern antibiotics, but since the 1990s rearing its head again in many places, encouraged by poverty, AIDS, and resistance to antibiotics. We see TB as largely a human disease, but it also occurs in many animals, significantly in cattle.

In the old days cattle were a prime source of TB in humans, and 40% of cows were infected, and the cause of what was known as 'milkmaid's disease'. It was a major concern, then, when eradication campaigns against TB in cattle in Europe were doing spectacularly well during the middle of the twentieth century, but failed to make an impact in south-west England and in Ireland. In Britain as a whole only 0.006% of cattle now have the disease, but in the south-west of the country as many as 0.04% suffer from it. Herd after herd is diagnosed as carrying the disease, and there are 200–300 herd 'breakdowns' per year. Thousands of cows are slaughtered. The disease in cattle in south-west England has persisted without let-up, throughout the 1970s, 1980s and 1990s. The cause is badgers (Neal & Cheeseman 1996).

Areas where bovine TB is found most often in England are almost invariably areas where densities of the Eurasian badger are amongst the highest in the country, and probably in Europe. In some parts there may be as many as 30 badgers per square kilometre. Badgers get TB, and they may die of it, but they can carry it around with them for sometimes many years before they expire. Bacilli are distributed by badgers throughout their range, especially in urine, but also in faeces and saliva. The big problem is that badgers spend a lot of time exactly where the cattle are, in pastures. Cattle get infected mostly by breathing in the bacteria, presumably when they are grazing.

In these high-rainfall areas in the south-west of England and in Ireland badgers feed mostly on earthworms that they catch on the surface, mostly on short-grass pasture. Cattle farming has created the ultimate habitat for badgers, with a glut of food. Hence the improbable sounding densities of badgers. In many of these places 10–20% of badgers may carry TB, and when following the animals with radio-collars I have watched them picking up earthworms right around sleeping cows in their pastures. Clearly, through the badgers, the cattle are highly exposed to *Mycobacterium bovis*. So far, the evidence suggests that badgers are the only wild animals that are significantly involved in the TB scenario.

The reaction of the Ministry of Agriculture animal health officials was predictable: kill the badgers (Dunnet *et al.* 1986). Kill they did, first by gassing, then by trapping, thousands upon thousands of animals. It started in 1975, and now, in the early twenty-first century, it still has not made any substantial difference to the prevalence of TB in either cattle or badgers. Almost everyone is agreed that badgers are the culprits in carrying TB (though they may have acquired it from cattle in the first place, and it is just possible that some other species of animals may also be involved), but ecologists are convinced that the official reaction has been inept.

There are several possible reasons why the killing is ineffectual, such as officials removing badgers from the wrong dens (the spectacular 'main setts', whilst inferior badgers tend to sleep in the inconspicuous 'outlier' dens). Also, badgers may move about more once the removals start. In the meantime, the killing continues. I am convinced that it would be more sensible, although difficult, to reduce the reason why badgers come to the pastures in the first place, namely, the presence of earthworms, because we appear to be encouraging badgers by our farming methods. Or we should attack the problem with oral vaccines as in rabies outbreaks (and this is the method now being researched), or address the problem of badger access to cattle.

Whatever the details of the clash between livestock farming and badgers over bovine TB, it is enormously expensive, and the total costs have soared well beyond millions of pounds in England alone. Moreover, as with rabies, the issue of losing livestock is one of personal loss for a farmer. Having part of your herd of prize cattle taken away for slaughter is harrowing, and it is a loss that cannot be measured only in cash.

There are other health threats to livestock in which carnivores may play minor, or sometimes significant roles. Brucellosis is known as 'spontaneous abortion' in cattle, it is common for instance in many African countries, and veterinarians have also found it in many African carnivores. But how frequently cattle actually pick up the disease from wild animals is not known.

In my Scottish otter study area an aborted otter foetus was found, and the cause of the abortion was diagnosed as a bacterium, *Plesiomonas shigelloides* (Weber & Roberts 1990), that also occurs in man, cats and dogs, causing diarrhoea. This was an isolated case, and we do not know how transmission took place; we cannot decide whether otters may pick the disease up from domestic animals, or vice versa. But potentially, these and other such maladies in wild carnivores are a hazard to all livestock that they come in contact with.

A wildlife ecologist pointed out to me recently that I am taking a human-centred viewpoint of such cases, that I do not sufficiently take into account the fact that we ourselves or our domestic animals may transmit diseases to the wild carnivores, and that in my discussion I take sides against the latter. This, as I will elaborate later, I do here on purpose. I am as fervent a supporter of these wild animals as anybody, but we need to understand what people have against them, and why instinctive reactions are often negative. Only when I articulate the negative (as well as the positive) aspects of carnivores as people perceive them, can we understand the emotions and reactions against the animals, and later protect them against ourselves. As will become clear in later chapters, conservation is one of my aims.

OTHER DAMAGE FROM CARNIVORES

There are some other, minor crimes committed by the Carnivora, and I want to mention a few of these for the sake of completeness. In fact, the range of different sins against human existence perpetrated by wild carnivores is almost as long as the number of species, and I can do little more than select a few of the larger escapades. One case that I must admit to finding rather amusing (but it must be maddening for the victims) is that of the beech marten.

Like a human criminal, the beech marten targets the well off in society in Central Europe. It goes for one of people's soft spots, the wiring in their parked cars, often the really classy vehicles in the sub-urbs of southern Germany and northern Switzerland. The martens slip in under the bonnet of a nice new BMW, they sleep on a warm engine, and they bite, claw and thereby strip the cables and insulation, causing mayhem on a substantial scale. The habit started around 1980, but now several thousand cases are well documented, with the damage already amounting to many tens of thousands of dollars. What is more, the habit is spreading (Kugelschafter *et al.* 1993), and we are still quite in the dark about the motivation for this destruction.

In North America raccoons and skunks often take up residence in houses, and they can do considerable damage (apart from being a health hazard). There are thriving businesses that specialize in removing these animals, so at least somebody is profiting.

Honey badger (ratel)

Close to my home in Britain, badgers often cause damage to cereals, especially in oat fields where day after day the animals may flatten and eat the crop. Elsewhere, maize is one of the badger's favourite targets (Kruuk 1989). They also annoy farmers by digging large holes in the fields or under country roads, which are ideal places for getting your vehicle stuck.

Another badger, the African honey badger or ratel, has often been cursed by bee-keepers when it raids their hives that they put out in the acacia woodlands. It often totally destroys the hive in the process, and eats the grubs and the honey, like bears do elsewhere. Once I visited a tourist lodge (in Mikumi National Park, Tanzania), where just 2 days earlier a couple of honey badgers had torn the large, metal freezer apart in the kitchen. They left the place as if there had been an explosion.

Carnivores provide many other sources of annoyance. These include otters ripping fishing nets apart, which I saw done by spotted-necked otters in Tanzania, and by their smooth-coated relatives in Thailand. For scavenging nuisance around people's houses, especially in Africa, hyaenas take the prize, as I know to my personal cost. Any domestic animals, food, shoes and other pieces of clothing are taken and eaten by hyaenas, and one might be forgiven for thinking that the animals have no limits to their digestive abilities. One of my zoologist neighbours in the Serengeti left an antelope's head in a bucket of formalin solution out on his verandah. The next morning it had gone, and hyaena tracks told the tale. The robber must have dipped its whole head into the formalin in order to get the booty, but history never related what its later fate was.

In the ultimate and final indignity to our self-esteem, hyaenas unearth graves or they take corpses of people who have died in the open. This happens in Africa and in many parts of Asia, and wolves are known to do it in other areas. Perhaps there are times when such behaviour may provide a welcome service, and several African tribes (including the Masai) in many places still put out their dead deliberately for disposal by hyaenas. But however useful it may be in some circumstances, in general it is an activity that is almost designed to evoke our deepest loathing. These wild animals deprive us of our very last, quiet resting place.

Whatever one's opinion and condemnation of these carnivore activities, they certainly add some excitement to our life, and I have heard many a tale about the carnivore's crimes told with an admiring tone in the accuser's voice. I believe that in general, urban spectators are quite prepared to accept that these predators do damage to game and live-stock, but feel that one should live with it. It is only when large-scale damage is done that people really suffer. People at the forefront, like farmers and gamekeepers, are far less sanguine about carnivores. It is there that the competition with wild animals bites, moulding man's anti-predator behaviour.

Sabre-tooth tiger

6

History of a conflict
Carnivores and the first hominids

Much as I may like carnivores myself, in previous chapters I had to emphasize the fact that people today may still be personally threatened by predators, and that we also lose to them when we rear livestock or chase game. Fortunately, at present, the general public (at least those of us living in rather urban, western society) are generally rather tolerant about these threats, losses and dangers, despite the problems that they are causing for sections of our populations. People in the West, and in most other countries of the world, are generally fascinated by wild carnivores, and most countries have now enacted legislation aimed at keeping many of these high-profile animals around. But that we and our livestock are often targeted by these same predators is not in doubt.

One obvious question is whether our predicament as a prey species is a hangover from our very early prehistory, or whether it is a more recent development. How did the conflict with carnivores evolve, did new predators arrive after us or did we and our predators evolve simultaneously, or did our species emerge in a world already full of enemies? Insights into the prehistory and evolution of our species in the face of pressures from the carnivores would prove useful. Not surprisingly however, actual evidence of past interaction between them and us is scarce and hard to come by.

PREDATION ON EARLY MANKIND

The fact that there are few data on the prehistory of predation on mankind is hardly surprising, because predators and scavengers are remarkably efficient. To demonstrate such efficiency, take a kill in the

Serengeti, an everyday example of many that I watched when I lived there.

A pack of spotted hyaenas chase a zebra family at night, and by biting at her legs and flanks they manage to slow down one of the mares. The stallion attacks the hyaenas, but there is little he can do against a dozen of the tormentors, and within minutes of the first bite the mare is down, whilst the rest of her family runs on. Hyaenas tear away at the flesh, and more of them join. A total of 34 hyaenas eat from the victim, and 40 minutes after being pulled down there is nothing left of the zebra, just a large, dark stain on the grass and a steaming heap of stomach contents. Somewhere a hyaena will be chewing on a jaw, reducing it to no more than a set of teeth. But the rest of the prey, including all the large bones, will be digested, totally.

Events such as this are a common occurrence. Different areas, different predators and prey, but the trend is the same everywhere: usually everything is eaten, if not by the predator itself then by scavengers coming afterwards. Such observations demonstrate that the chance of any prey item landing itself into the fossil record (e.g. by getting neatly piled up in a cave) is extremely slim. Nevertheless, bones in caves are most of what we have as evidence from the past. They are not much use for finding out exactly what happened, and almost inevitably, the most we can deduce is who was contemporaneous with whom. With luck we can sometimes suggest which species of animal chewed the bones.

It was an unusually lucky break, therefore, when in the 1970s the South African scientist Bob Brain came across the skull of an early hominid child in a cave at Swartkrans, with clear indentations by a set of canines, most likely those of a leopard (Brain 1981). Of course, this did not mean that without any doubt a leopard had killed the little victim a million years ago, but from the size of the tooth marks it is highly likely that it was a leopard that dragged the corpse, holding it by the head, as leopards often do. Because leopards do not scavenge that much, it is probable that the same animal had also killed the child. This incident is the earliest evidence we have of predation on man.

The cradle of our species, or at least one of the cradles, stood in East Africa. One of the most famous sites is Koobi Fora in Kenya, on the eastern shore of Lake Turkana (formerly Lake Rudolf). It is dry bush country, an almost endless, wild and blistering hot semi-desert area of Northern Kenya, close to the border with Ethiopia. It is cut through by dry river beds and by rugged mountain ranges next to a lake as big as a sea. When you are there you cannot help but be awed by the domination of the African bush and its inhabitants, although you are no more threatened there than in an average city. But predators are there, people are very few and far between, and you are a potential prey for several species of carnivore. Predatory violence is commonplace, and when I was there on my first day I had a terrific fright when I almost walked into two lions next to a carcass of a large crocodile that they had killed the previous night.

Lions, hyaenas and leopards were the reason for my presence at Koobi Fora, where I was working for UNEP. It is nomad country, territory of the Boran, Gabra, Turkana and other tribes with romantic sounding names, people who graze their camels, goats and other livestock. The nomads are totally dependent on their animals, but lions, hyaenas and others are a constant menace to their existence.

One evening when I was there we went hyaena counting, for a survey that I was carrying out to assess numbers. I had a method: I put a large loudspeaker on the roof of my Land-Rover and broadcast calls of hyaenas squabbling over a kill. Hyaenas came from over 2 miles away and I could count them, recognize individuals, and get an estimate of a population. My small children sat in the back and we all stared into the black African night around us, the air rent by the hideous howls from above our heads. Four large hyaenas excitedly circled us, just a few feet away. They were very tame, and one is easily fooled by that. Suddenly there was a yell from the back of the car: a large male lion was right behind us, attracted by the hyaena noises which to him usually meant food.

My children saw it as a life-threatening incident, and the sudden appearance of a lion from the darkness, to within inches of their faces was terrifying for them. It reminded me again of the necessity to look over my shoulder when out in the bush, and later I realized that it was especially significant in that particular place.

Koobi Fora is one of the most important sources in the world for our knowledge of human evolution. Here, just in the same place of that lion threat to my children on that day, fossils were scattered on the surface. I could see the remains of lots of different mammals even from my car window. This was where fossil bones of the earliest of man, of *Australopithecus* and *Homo* were found. This was where the palaeo-anthropologist Richard Leakey established his camp. It was mostly from here, from good fossil evidence, that he built up a picture of what early humans were like, and what they had to contend with. What intrigued me was whether large carnivores were part of the scenario, at the time when early mankind evolved in this very same spot, just like today.

In Koobi Fora, amongst the fossils, one feels very close to the past, and perhaps even more than elsewhere that the past is only a short while ago. It is, indeed, amazing how fast our human society has evolved. Thinking back over my own lifetime, one generation is so short: it is the time between me and my parents or between me and my children. Yet, if an average generation lasts 25 years, then it is only 80 generations since the time of Christ, and 800 generations ago we were still in the age of cave paintings in Europe. *Homo sapiens* has been around for only 8000 generations. This is the brief period in which we have had such a profound effect on the planet, most of it over just the last few generations.

The earliest known hominid species date from about 4 million years ago (4 mya), although there probably were primates walking bipedally at least twice as long ago (Leakey 1994). In East Africa, at about 4 mya, species of *Australopithecus* and *Homo* shared and maybe competed with each other for resources. About 2.5 mya *Homo* began to show a dramatic increase in brain size and the first tools appeared. About 2 mya *Homo erectus* expanded into Eurasia, and tool use became widespread. The first human use of fire occurred 0.7 mya, or 700 thousand years ago, and it is now estimated that the first fossils that are definitely of our own species, *Homo sapiens*, are only about 200 000 years old. The oldest cave paintings, in Africa and Europe, are from about 30 000 years ago.

The habitats in which early *Homo* operated have been discussed in several studies, and a recent analysis shows that earlier hominids such as *Australopithecus* probably lived in wooded and well-watered regions, whilst *Homo* was the first hominid genus to use open, sparsely wooded and arid grasslands as its main habitat (Leakey & Lewin 1979; Reed 1997). Many other studies have also recognized the open savannah as the landscape where *Homo* evolved, which has never surprised me. Regions such as the Serengeti, with large trees and small rocky hills scattered over gently rolling pastures are beautiful to us, and in country parks we try to recreate such landscapes. It is exactly the scene where instinctively I feel most at home; somehow it seems not out of place that this should be the habitat of the origin of people.

But it was also a natural habitat for many other mammals. When I summarize the ecological niche of early *Homo* as it has emerged from all these studies, as the niche of a gregarious hunter of grazing animals in open savannah country, I could also be talking about wild dogs, spotted hyaenas or lions: their ecological niche is rather similar to that of hunting man. Early people beat themselves a path that was bound to lead them into direct conflict with several well-established, well-adapted and well-armed predators, all doing the same thing in the same habitat.

EARLY CARNIVORES

In Koobi Fora, the Leakey teams found fossils of several predators that threaten people now, such as spotted and striped hyaenas and lions, and they lived at the same time as the earliest hominids. There were also others, long since gone. Many of them were large species, which

Spotted hyaenas

would have been of more than passing significance to a person when met on a dark night or in a lonely spot.

At the time of the rise of mankind there were various species of false sabre-tooth cats, *Dinofelis*, and 'proper' sabre-tooth cats, for instance several leopard-sized *Megantereon* species, and *Homotherium* and *Machairodus*, which were at least as big as a lion. Additionally there were large species of hyaenids such as *Percrocuta* and *Euryboas* (Leakey 1976). Judging from the fossils they must have been formidable animals, some much bigger than their relatives of today, and all living cheek by jowl with people, some 3.5 to 1.5 mya. There were also many smaller predators sharing the habitat with humans. These included otters, honey badgers, mongooses, civets, and for almost all these groups palaeontologists have recorded a larger number of species than there are now in East Africa. This means that at the time of the introduction of humans into the ecosystem there were many more different species of carnivore around than at present.

No doubt in future some of the dates of human evolution will be moved back a bit, with new finds of fossils. But there will never be any doubt that the Carnivora were present long, long before hominids arrived on the scene. The ancestral carnivore family was that of the Miacidae, small predators of a size and shape comparable to that of a genet cat, now extinct. Between 55 and 40 million years ago many other families branched off, including several that are still here. At 10 mya all the present-day carnivore families were in place, with many of the same species as today (Bininda-Emonds *et al.* 1999; Van Valkenburgh 1999).

One thing on which palaeontologists agree is that animal evolutionary history over this last 40 million year period, and even before then, is hugely complicated. Our knowledge is the result of studies by many palaeontologists (e.g. Savage 1977, 1978; Brain 1981; Turner 1985, 1990; and reviews in Martin 1989, Macdonald 1992, Hunt 1996, Werdelin 1996 and Van Valkenburgh 1999), and I can only provide a brief précis. The story involved a plethora of different carnivorous mammals, in a pattern that repeated itself again and again: on at least seven occasions an entire group (family, superfamily or even an order) of these mammals came on the scene and later disappeared again, each monopolizing the predator fauna for a time, then being replaced by another group. One such major changeover occurred in the beginning when the Order Carnivora appeared, replacing the Order Creodonta.

Fossil finds showed that in those early, heady days of the family Miacidae there were already many other predators, different from the Carnivora and coexisting with them. Often these other species were large ones, including the now extinct order Creodonta, with families such as the Hyaenodontidae. Apart from the creodonts there were also various large marsupial predators such as the *Thylacoleo* or pouch lion. All these have become extinct, whilst the Carnivora blossomed.

Thus, fossil analyses show proper carnivorous feeding and predation in several independent lines, at least twice amongst the marsupials (in the borhyaenids in South America and some dasyurids in Australia) and several times amongst placental mammals (including the now extinct creodonts, and the surviving Carnivora). The big carnivorous expansion in the late Miocene and early Pliocene coincided with an explosion in the evolution of flowering plants and grasses, and it is easy to see why. The evolution of entirely new types of plant resulted in a huge floral diversification, creating new habitats, including savannah-type vegetation. This, in turn, enabled the evolution of an enormous diversity of herbivores, and therefore of potential prey to predators. It laid the table for the carnivorous fauna.

The last creodonts lived some 8 mya, so they did not see hominids. But the last really big marsupial carnivores lived up to 2 mya, when *Homo* was already well established. Of the smaller marsupial carnivores just a few small dasyurids are still hanging on now in Australia, such as the Tasmanian devil and the quoll. Whilst the other predators slowly disappeared from the scene, the Order Carnivora thrived and evolved into a multitude of different families, genera and species.

Despite this diversification, most of the Carnivora have also become extinct. We still have a rich party of species, but it was a riot of biodiversity in the past. The most recent summary of what we have lost shows that there were at least 481 genera of Carnivora, of which 352 (73%) are now extinct, and only 129 (27%) still alive today. In terms of species the numbers of extinctions are even more daunting. A number of complete Carnivora families have gone the same way as the creodonts and large marsupial predators (Savage 1978; Van Valkenburgh 1999), and they are totally extinct now.

Why the creodonts disappeared and carnivores thrived is still a mystery (Van Valkenburgh 1999). They were rather closely related, the skeletal remains were not that different (the variation between bones of creodonts and of carnivores was not much greater than the variation within these groups, although of course there is more to an animal than its skeleton), and they overlapped in the same areas for a considerable period. It is quite likely that the two groups competed for resources, but probably we shall never know what selection pressures were operating that favoured the carnivores, and did away with the creodonts. One

mechanism that could be important is that groups or families of creodonts evolved and slowly changed in character, starting as generalists,
but gradually producing highly specialized predators, like the present-
day cheetah and the wild dog amongst the Carnivora. By their nature
these specialists will be more susceptible to environmental change, and
therefore, more likely to become extinct – making way for another wave
of generalists (Van Valkenburgh 1999). If this were a general trend, then
families of Carnivora could also be headed in the same direction.

Within the cat family the present-day large species, often collectively called the pantherines, evolved between about 4.2 and 1.5 mya
(Bininda-Emonds *et al.* 1999). This means that they are relatively recent,
but at the same time they are highly specialized. As Bob Savage put it:
'The felids can truly be regarded as the acme of carnivore evolution;
though today limited to a few genera, their many species are widely
distributed. In almost all their skeletal elements, in their senses and
dental apparatus, they represent the ultimate in carnivore achievement'
(Savage 1978). This felid evolution took place more or less contempor-
aneously with that of the hominids. It included the evolution in the
large cats of the habit of preying on hominids, and of competing with
them for the same prey species.

The evolution of mankind's anti-predator behaviour against the
large cats, therefore, was part of an arms race right from the beginning.
Formidable felids such as the North American sabre-tooth cat *Smilodon*
did not emerge until as recently as 2 million years ago, at the same time
as species of *Homo*. *Smilodon* did not last long, though, as it disappeared
again about 9.5 thousand years ago; but it was around for long enough
for it to have seen a good deal of primitive man, and vice versa. Most of
the other old sabre-tooth cats, the 'palaeo-sabres', evolved 40 mya and
were totally extinct by 6 mya, well before man arrived.

As an aside, the sabre-tooth felids still pose one of the all-time
fossil predation mysteries. Large canine teeth are used by today's carnivores for killing prey, as well as for social purposes such as fighting
opponents over territorial claims. But why did the early cats have those
extra-large sabre-teeth, huge flat daggers which were seemingly far too
large for any jaw? Were they used to kill extra-large prey, or for opening
carcasses, or what? In fossil assemblages it was always the very largest
ones, the top predators, which sported sabre-teeth. So far nobody has
come up with a likely explanation for the use of sabre-teeth in the
acquisition of food. The fragile, sharp weapons, often with serrated in-
ner edges, must have been quite useless against thick skin or on large
bodies, because the gape of the owner is insufficient to use the canines
effectively.

Yet sabre-teeth were obviously eminently useful and effective weapons, otherwise they would not have evolved independently in various different orders and families, and in both sexes. Interestingly, the carriers of sabre-teeth also died out again in all these cases. The chances are that sabre-teeth made use of some Achilles heel in their prey (of which we have no evidence at all today), but that in response the prey species evolved a means of protection. It was an arms race that was eventually lost by the sabre-teeth carriers, but we do not know who conquered, and why.

Other carnivores that were large enough to prey on stone-age 'man the hunter' were some of the canids, hyaenas and bears. The wolf is a product of extensive branching and rapid evolution amongst the dogs, some 2.5 mya (Bininda-Emonds *et al.* 1999), just like the large felids, and again, at the same time as we evolved. Another significant group of large canids were the borophagines (the hyaena-dogs), which evolved quite a bit earlier at 20 mya, and disappeared in the sink of extinction about 1.5 mya, well within human time on earth.

Most of the speciation amongst hyaenids took place about 10 mya, and at the beginning of man's evolution there were at least nine species in Africa alone, several of them larger than the three present ones. But the family also ranged widely over Europe and Asia, and we can justifiably surmise that there were several hunting species of hyaena constituting a danger to early man on all these continents, including the spotted hyaena which is still a potential threat today.

The last family with really large members is that of the bears. Most of them have evolved very recently, some 5.7 to 1 million years ago. Amongst the extinct species were the huge, formidable cave bears, which must have been quite common in some places. In one Austrian cave alone bones of some 50 000 individuals were found (Kurten 1968). The cave bears left the scene well after the arrival of mankind, about 10 000 years ago, a period equivalent to only 400 generations of people.

All the evidence that we have suggests that with so many more species of predators around than nowadays, *Homo* is likely to have been much more threatened in the early days of its evolution than recently. Our ancestors were entirely within the size range of prey species, and their hunting–gathering way of life probably exposed them to large carnivores much more than the present-day life of agriculturalists. It is not difficult to picture a scenario. The Japanese scientist Tsukahara and his team showed how present-day lions are serious predators of chimpanzees in Tanzania (Tsukahara 1993), and in prehistory just as now, there must have been the party of foraging primates, scattered through the dry, open woodland, and the pride of lions or other carnivores, hidden in formation and closing in, stalking, rushing, killing in broad daylight or at night. It still happens today, and it is under those conditions that our primate anti-predator systems evolved, as I will discuss later. Man must have been a welcome addition to the prey spectrum of many carnivores, and there are no reasons to assume that maneating was not a normal aspect of day-to-day predation during the Pliocene and Pleistocene.

HUNTING AND SCAVENGING IN EARLY HOMINID EVOLUTION

Not only were the large predators an immediate and direct threat to our early survival, but there was an additional, negative aspect. When hominids evolved, they came in as direct competitors with the other hunting species. There is good evidence for early man as a hunter, in the same ecosystems where the large carnivores operated, evidence for instance from caves in South Africa. Fossils showed that neolithic mankind had abundant access to prey, especially to ungulates of all sizes. There were many signs of butchering, and at least some bones that were found were of animals such as buffalo which had been hunted rather than scavenged, as arrow tips were found stuck in the bones (Milo 1998). Man was a predator, as much as a lion, a wolf or a cave bear. Carnivores, therefore, were our direct competitors.

However, early people were not only adventurous hunters. Researchers of our evolutionary history found that scavenging was also a major form of food acquisition. For instance, hominid bones from Olduvai in the Serengeti, 2–1.7 million years old, were clearly associated with remains of large ungulates. Markings on some of the animal bones indicated that they had been scavenged by man *after* carnivore predators had chewed them and presumably eaten their fill (Shipman 1986). These bones, together with data from many other sites, demonstrate the carnivorous leanings of humans from a very early stage in their evolution, and show human exploitation of animal populations by scavenging as well as hunting. The observations imply that to carnivores the newly arrived *Australopithecus* and *Homo* were potential competitors (by scavenging from the carnivores as well as by hunting the same ungulates) as well as being potential prey.

George Schaller and his colleague Gordon Lowther demonstrated how such competition might work (Schaller & Lowther 1969). By walking about in the Serengeti for several days on end, they found it perfectly possible for unarmed human scavengers to survive in that habitat. They took their cue from descending vultures, and, by finding carrion

Tiger

and chasing lions and hyaenas away from their kills by approaching them in full view and shouting, they could easily steal enough meat to survive. One has to keep in mind that present-day lions are conditioned to people being armed and dangerous (so probably they can more easily be chased away now), but, even without lions, the two scientists would have found enough carrion. It is probably safe to assume that if they could scavenge enough, then Stone-Age man would have been better at it, and would have foraged extensively that way.

As far as actual hunting is concerned, there are interesting parallels to be drawn between hunting man and several species of large carnivores. This holds not only for their prey selection, where man concentrated on large herbivores just like the large cats and hyaenas and dogs, but also for other similarities in hunting strategies, as suggested by present-day methods used by several African hunting tribes, which involve the stalking and sometimes the driving of prey.

There are also similarities in their social organization. As we have seen, species such as the wolf, spotted hyaena, lion and others hold group territories, in societies that are often much larger than the individual groups (packs) in which they forage. They live in fission–fusion societies (Chapter 2). It is not difficult to see that there are many parallels between the organizations of gregarious carnivores and such social structures in 'primitive' people's hunting societies, as well as in societies of chimpanzees. The similarities are interesting, not just for their own sake or for demonstrating the ecological roots of mankind, but they also emphasize the close competition that must have developed between us and the hunting animals. The more similar their behavioural foraging techniques and social background, the more likely species are to compete with each other.

It seems likely, therefore, that competition for resources was present, even long before people started keeping livestock, now about 9 millenia ago. That was an additional development leading to conflict, and although we have no direct, hard evidence that predators targeted our early animal husbandry, there should be little doubt that mankind's first attempts at keeping goats, sheep and cattle, attracted more lethal attention from carnivores than they do now. On all scores, ecological proximity, predation and competition for resources between people and carnivores are likely to have coloured the relationship over many millennia, and only over the last few generations has this relaxed somewhat. This happened because mankind abandoned much of its hunting tradition, eradicated some of the predators and moved livestock out of harm's way. Things may have eased somewhat, therefore, but there is no doubt about the long-standing evolutionary roots of our present-day antagonism.

CARNIVORE EXTINCTIONS AND *HOMO*

After the long struggle between our species and the Carnivora, on many fronts and with many battles fought in different ways, it is glaringly obvious that mankind has come out on top. Are we, then, the cause of the many extinctions amongst those who always were our foes?

This is a question that is often asked (e.g. Walker 1984; Vrba 1985, 1988; Turner 1990), and answers are far from straightforward. One notes that many carnivore species extinctions occurred around 4 mya, just at the time that the first hominids arrived on the scene, and many other species followed them into oblivion over the next 2 million years. But equally clearly, many other extinctions occurred well before humans arrived. Moreover, we also have to bear in mind that the emergence of people was not the only event that changed the face of the earth at and after that time. There were drastic changes in climate, too, for instance around 3.2 mya, 2.4 mya and 0.8 mya. The Pleistocene period started at 0.8 mya with a massive climatic shift, involving considerably higher temperatures. This heralded the expansion into Eurasia of several African species such as the lion, leopard, spotted hyaena and perhaps the major (although not the first) movement out of Africa of our own genus *Homo* (Turner 1990).

If such major dispersal events occurred in conjunction with climatic changes, it is also likely that extinctions could have been initiated. For instance, the palaeontologist Alan Turner remarked on the climatic events around 2.4 mya, which coincided with 'a major faunal turn-over that reflects the appearance of modern, cursorial, grass-adapted forms [of ungulate]. The effects of that change... are likely to have had a major impact on the archaic, machoirodont [sabre-tooth] cats, who may have found prey increasingly difficult to catch' (Turner 1990). The point is that of many of the extinctions of the earlier carnivores, just before or just after the first hominids came onto the scene, we will never be able to say *mea culpa* with any conviction, because the only evidence we have is that of coincidence in time, of one species increasing and of others disappearing. Perhaps people had nothing to do with these disappearances. Perhaps it was all our fault, perhaps it was also due to other factors: environmental change may have rendered species more vulnerable to our onslaught, and to competition.

However, for the most recent extinctions the actual mechanism has been well documented, and here the guilt of mankind is in no doubt. For instance, the largest recent marsupial carnivore, the Tasmanian wolf or thylacine, was still around in Tasmania when I was born. It was shot and poisoned into oblivion by sheep farmers in the 1920s and 1930s (Paddle 2000). The North American sea mink was obliterated in the late nineteenth century for its fur, and the sea otter almost shared its fate for the same reason, saved by the bell, but only just. The 'wolf' of the Falkland Islands, which is related to the South American fox-like canids, was still there when Charles Darwin visited in the first half of the nineteenth century, but it preyed on sheep, and because it also carried a useful pelt, it has existed since the 1880s only in museums.

The plight of the giant panda and the tiger are only too well known, and I will discuss them in more detail in later chapters. Several carnivores have been totally eradicated from Britain, such as the brown bear and the wolf, whilst the wildcat, polecat and pine marten have only just managed to hang on here. One can go on with long lists from all over the world where we know the causal agent to be mankind, although perhaps some species in a few places were also doomed in some other way. Undoubtedly, the vast majority of extinctions in the last few centuries were induced by humans.

We wrought a terrible revenge on the beautiful animals that bothered us, and people exploited them, and harvested furs until the end of the goose that laid the golden eggs. I can only hope that, now we have done our worst, we can make sure that no more extinctions will follow. But I must admit that I am not optimistic.

Producers and products

7

What is the use?
Carnivores as food, for medicine, perfumes, sport, tourism and the fur trade

Previous chapters showed carnivores as a dangerous and expensive nuisance, but it would go against the grain to leave it at that. They are highly attractive animals, and if that is not enough, wild carnivores can also be tremendously useful to us. In this chapter I will discuss some of their direct, material benefits – often more of historical significance than of present-day importance, but nevertheless far-reaching aspects of our ecological relationship with carnivores.

Some time ago I gave a talk here in my Scottish village, and when it came to the conservation of otters, a somewhat grumpy farmer asked 'what is the use of them to me?' He was, of course, implying that otters and their conservation are a waste of time. To many people, like me, otters are of use, simply because we are thrilled if we see them, and even if we do not actually come across an otter we enjoy the knowledge that we might, and that they are there in the river or the lake. People are even prepared to pay for this pleasure. Nevertheless, the farmer's utilitarian question is central to society's approach to the environment. Can we eat them, use their products, or do they provide a sport or service? In the end it is this question that will determine the survival of many of these species.

CARNIVORES AS FOOD

In modern western society we do not eat carnivores any more; even the idea is rather repulsive. But it is only less than a century ago that several different carnivore species were considered delicacies in many western countries, and just out of curiosity I myself sampled some.

One early morning I witnessed a battle between two male lions in the Serengeti. The younger intruder got the worst of it and died about half an hour after the victorious territory owner left the scene. It gave me the opportunity to take the hind legs of the victim, and that evening we had a magnificent barbecue with my colleagues and their families. The meat was beautifully tender with the flavour of a somewhat gamey escalope of veal, and it was excellent. Everybody loved it, and no one guessed what the steaks were. But afterwards many were horrified when I told them, and one of the scientists was almost sick. The story has two points. Firstly, the meat of at least some of the carnivores has a lovely flavour. Secondly, the very thought of eating a carnivorous animal can be very off-putting.

Last year, in a remote Belarusian village, a recently killed lynx was presented to our small research group by a local poacher, who just wanted to have the skin. The meat was quite excellent, with a delicate, smooth taste. I also remember from my younger years during and even well after World War II that cats in Holland were nicknamed 'roof hares'. They were regularly sold by the locals and even by butchers as hares (skinned, of course, and with the head and paws cut off). Especially just before Christmas we had to keep our cat indoors, because hare was a traditional and favourite Christmas treat.

Bears used to be eaten in Europe and North America, and even in the 1960s and 1970s there were restaurants in Germany offering bear paws as a speciality. One of the classic English–American cookery books (*The Joy of Cooking*, Rombauer & Becker 1963) also has a recipe for bear, which states specifically that all bears except the black bear are edible. But never in the history of these western countries have bears been so popular for food as they were and still are in eastern Asia. They are often eaten in Thailand, Vietnam, Cambodia, China, Korea and Japan, despite official protection in several of these countries. Vietnam, Cambodia and Thailand see throngs of rich tourists from Taiwan, Korea and Japan visiting restaurants especially to eat bear, with bear paws being the favoured parts. Front paws are better than hind paws, it is alleged, because the animals lick them so much. Often the animals are beaten and tortured before being killed, as this increases their adrenalin production and people say that this 'fear juice' makes the meat especially tasty and tender (http://www.earthtrust.org/bear.html). I find it almost impossible to comment on this.

In my own limited experience, I did not find all carnivores good to eat. I once fried steaks of the Eurasian badger and they were terrible: very greasy and with an extremely strong game flavour. However, I later learned from a German recipe book that this applies only to older animals: a young badger is described as having a very good, subtle taste, especially in autumn (Horn 1964), and certainly in the nineteenth and early twentieth centuries they were frequently eaten.

In the not so distant past, during times of fasting for Roman Catholics (especially Fridays and the six weeks before Easter), the only animal food allowed was fish, a rule that the Vatican abandoned for ordinary folk only in the 1960s. As an interesting exception during these fasting days, however, otters and beavers were considered to be honorary fish, because of their aquatic habits, and there are several recipes for otter which were popular especially in Germany and Austria (see Box 7.1).

Box 7.1 Recipes

Badger (from Austria)
After skinning, all fat should be removed, the meat lightly salted and braised with some onions, for about one hour. Add a few juniper berries, and the badger should be served with a wine sauce and sour cream. Alternatively, one can prepare 'jugged badger' by cutting the meat in chunks, marinading it for 2–3 days, boiling with a tenderizer until soft, then adding a laurel leaf, peppercorns, salt, red wine, sour cream and stirring in some apple sauce.
(after Siebold 1959 in *Die Wildküche*)

Bear (from the USA)
Remove all the fat immediately, because it quickly goes rancid. Marinade the meat for at least 1 day, then cook it as one does any beef pot roast or stew. Meat of a bear cub needs to cook for at least $2^1/_2$ hours, that of an adult for 3–4 hours.
(after Rombauer & Becker 1963 in *The Joy of Cooking*)

Raccoon (from the USA)
Soak the meat overnight in salt water, scrape off all the fat, then blanch it for an hour. Add some soda and cook for another 5 minutes, then wash it, put it in cold water and bring it to the boil, then simmer for 15 minutes. Finally stuff the raccoon with a bread dressing, bake in a 350 ° oven for 1 hour, and serve.
(after Rombauer & Becker 1963 in *The Joy of Cooking*)

Box 7.1 *cont*

Otter aux fines herbes (from Germany)
Add a mixture of herbs (including thyme, capers, basil, garlic),
and a few anchovies to water and some oil. Add chunks of salted
and peppered otter meat, simmer until soft, then add a glass of
wine and simmer for a few more minutes. Remove the meat,
continue simmering and when the sauce is suitably reduced, add
some flour and a vegetable stock, as well as a touch of vinegar
and lemon juice. Serve the meat in a deep dish, after covering it
with the hot sauce.
(after Jorga 1998 in *Der Lausitzer Wasserman Lebt*)

Dog caldereta (from the Philippines; for Saint's days or other
special celebrations)
Fry a chopped onion and a clove of crushed garlic until brown.
Add 1 kg of cubed dog meat. Stir and fry until brown, then add a
little water. Mix ½ kg of cubed potatoes, some chillies, tomato
sauce, dried paprika and add to the meat. Simmer until meat is
tender. Stir in a handful of crushed peanuts about 10 minutes
before serving.
(personal communication from Mrs A. Velasquez, Manila)

People in western countries are especially horrified by the idea of eating
dogs, for various reasons. I myself did not like the odour of the meat of
a freshly killed domestic dog, nor that of a hyaena or a fox. It put me off,
but people in South-East Asia and elsewhere, and the Sioux and other
Indian tribes in North America, thoroughly disagree with me on this
(as would have people in Europe in earlier days). Dogs are frequently on
the menu in South-East Asia, far more often than cats. They are often
sold in markets for meat, having been specially bred and fattened for
this purpose, and in Thailand and neighbouring countries one even
finds special dog abattoirs to provide the local delicacy (Corbett 1995).

Hunting tribes, such as the Sirione in Bolivia, regularly take small
carnivores such as raccoons, coatis and kinkajou, and these provide an
imortant source of meat (Stearman & Redford 1992). Also, the Hadza
people in northern Tanzania will catch and eat carnivores as often
as they are encountered, just like ungulates. Lions, hyaenas, leopards,
jackals and others are all consumed without prejudice (Woodburn
1968).

Nevertheless, even the staunchest animal advocates would agree
that, on the whole, carnivores nowadays are unimportant as a source of
protein, despite being considered in some countries as delicacies and

despite their use in the past in the west, and in eastern Asia, South America and Africa now. As a species, we rarely eat carnivores, and in many countries people avoid their consumption altogether. The reasons may be sheer irrational prejudice; people are put off by the idea, without being able to pinpoint why. Perhaps Western people dislike the thought of eating animals that have eaten others. But more likely there is a good, instinctive biological reason for the dislike, such as avoidance of lethal tapeworms and other parasites that may infect people as well as carnivores (although many of these also occur in pigs). I also think that our involvement with pets has something to do with it, in the same way as sympathy for horses makes British people abhor horse-meat. I will later return to this 'pet effect'.

MEDICINAL USES AND SCENT

However, the value of a wild animal to our society is not decided by its protein alone. Several carnivores are popular for their supposed medicinal properties, or at least they were in the past, although in not one single case has such medical value been demonstrated in western countries. But of course, it is more important what people believe, rather than what can objectively be shown to have an effect.

For instance, in Britain (especially in Wales) and many countries on the continent, badger fat is known as a cure against arthritis, although in Germany it was used for making soap. In South-East Asia dog meat cures fevers and stomach ailments (Corbett 1995), and to our medieval forebears in Europe, the wolf was a walking medicine chest.

Tiger

A live wolf drowned in oil was a certain cure for gout, and its heart drove epilepsy away. The fat protected joints against arthritis, and wolf blood overcame stomach pains. The animal's liver was a medicine against coughs, and carrying a wolf tooth calmed the demented and helped teething children (Gesner 1551–87).

In Tanzania, locals who knew that I was studying hyaenas asked me on several occasions if I could get them a piece of skin, or better still, hyaena heart or genitals. They told me that if you feed these bits to your livestock, then cattle would be protected against predation. What they did not readily admit, though, was that they also used it on their own persons, as people protect themselves against witchcraft by rubbing various hyaena substances into cuts in the arm (Kruuk 1975). It was not talked about, because witchcraft is illegal in Tanzania and carries a heavy sentence.

The use of the supposed or real healing properties of carnivores in Europe and Africa pales into insignificance compared with that in eastern Asiatic countries. There, body parts and various secretions from large carnivores are huge business, legal and illegal, involving thousands of poachers, hunters, farmers, pharmacists and doctors, and worth many millions of dollars per year.

Bears are amongst the worst exploited animals. This involves especially the Asiatic black bear, the sun bear and the sloth bear, and their exploitation and trade has been documented in detail on the Internet (http://www.earthtrust.org/bear.html). The animals are used for food to some extent, but by far their most valuable asset is their gall (gall bladders and bile), which is used for medical purposes.

Bear bile is used to reduce blood pressure. It is said to protect the liver and gall bladder, to dissolve gallstones, to reduce body heat, to detoxify the body and to treat coughs and asthma. It is also used to make shampoo, throat lozenges, cream for piles, herbal tea (sic), cough syrup and cosmetics. Many Taiwanese doctors prescribe it against several ailments, and there appears to be clinical Taiwanese research on the medical properties (although the data are not published in a public-access format) that shows bear bile to be an effective analgesic, anthelmintic, antipyretic and antiphlogistic (i.e. painkiller, de-wormer, fever killer and anti-inflammatory), and possibly but not certainly effective against convulsions, jaundice, ulcers and poor vision.

The use of bear bile as medicine therefore, has, official Chinese medical opinion behind it. The active ingredients are identified as ursodeoxycholyl-taurine, cholyl-taurine, chenodeoxycholyl and ursodeoxycholic acid, which can be substituted by various ingredients found in plants (including rhubarb) and in pig bile. However, it is generally believed that the original from a large animal's body is bound to be better, and also, perversely, the fact that conservationists stress the rarity of bears increases the demand for bear products: the rarer the better.

It is almost impossible to summarize the extent of the trade in bear gall bladders in Asia, but it is huge, and covers a vast area, with demand for many thousands of bears each year. It involves suffering on an enormous scale. Known details of the business are mostly of trade between countries, which for a commodity such as bear gall is only the very small tip of a large iceberg. Nevertheless, between 1978 and 1988, some 681 kg of sloth bear gall bladder was exported from India to Japan, involving between 8000 and 17 000 animals. Between 1979 and 1984 China exported between 7000 and 37 000 bear gall bladders to Japan, and more accurately we know that between 1988 and 1990, some 1051 kg of bear gall bladder, produced by around 10 000 bears, was exported from China to Japan.

These are only a few figures to indicate the scale of international trade, but the total numbers of bears involved locally in these countries may be an order of magnitude greater. Within China alone, the bile business is vast, despite official protection of wild bears since 1989: about three quarters of traditional medicine traders sell bile, and traditional medicine is applied everywhere, including government hospitals. In Taiwan, bear gall bladders sell for the equivalent of $30 per gram, and the average weight of a gall bladder is 114 grams.

In China there are hundreds of government-approved bear farms, which in 1993 were estimated to hold between 6000 and 8000 bears, and in 1994 there were estimated to be just below 10 000 bears (this does not account for the many illegal establishments). The inmates, Asiatic black bears, live in cages only slightly larger than they are themselves. From these farm animals bile is extracted without killing them, and in the more modern farms the bears have been inserted with stainless steel taps that are connected to the gall bladder, which enable the farmers to drain bile daily. More usually, bears wear abdominal shields ('vests'), covering a bag carried directly against the skin and a polythene tube from there to the gall bladder. Bile is collected daily from the bag, with the animal forced into a corner with bars, all with a great deal of growling and aggression. One bear earns the farmer about $2900 per year, with an output of 1–2.5 kg of bile.

The use of bears causes Chinese medicine to be strictly unacceptable to conservationists. Another compelling accusation against it is the use of bones and other body parts of tigers. There are a host of uses for tiger parts in China and elsewhere in the Far East, just as for bears, and just as there was for the wolves in medieval Europe. Tiger fat is supposed to make haemorrhoids disappear, the blood strengthens will-power, the testes protect against scrophula, the eyes improve vision, a suspended tiger nose induces the birth of boys, the whiskers cure toothache, the penis makes tigers of men, tiger bone wine is an aphrodisiac, the bones alleviate symptoms of rheumatism, and so on (Nowell & Jackson 1998).

There is no hard evidence for the efficacy of any of these 'medicines', but nevertheless a thriving black market has developed. Tiger bones sell for up to $300 per kilogram, and by obtaining tiger parts from everywhere in the species' range in Asia, this Chinese obsession has driven the tiger to a point where extinction is imminent. Recently, with tiger bones becoming very rare and with official protection measures more effective, bones from other species of large cat and other animals are being used as fake tiger parts.

Modern western medicine does not rely on animal products such as these. But it does use dogs, cats and ferrets in a different way, as experimental animals on which to test medicines, or for studying fundamental medical and biological problems. Rats and mice are used significantly more often in laboratory trials, but carnivores also feature. For example, of the 2.7 million laboratory procedures recorded in Britain in 1996, 85% involved rodents, 2.2% rabbits or ferrets, and only 0.4% (especially bred) dogs or cats. The percentage of experiments on carnivores was small, but it still included over 10 000 animals. Experiments on these animals for industrial purposes (such as by tobacco manufacturers) get a lot of adverse publicity, but they are very few in number compared with the medical procedures.

A totally different application of (what once was) a carnivore product is evident all around us: perfume. Over the centuries our social life has been much eased and enhanced by what Shakespeare referred to as 'the very uncleanly flux of a cat'. In Shakespeare's world a young man 'rubs himself with civet...the sweet youth's in love!', and King Lear's cry went up 'Give me an ounce of civet, to sweeten my imagination'. A person may briefly enter a room and leave again, but an hour later their fragrance may still linger. Scent or perfume is seriously important in our lives, and it clearly affects people's attractiveness. The civet cat was the usual producer of the vital ingredients of perfume, and the perfume industry was a very significant consumptive user of carnivores.

Civet cats belong to the viverrid family, and they have a long tradition of being exploited. The very strong smelling excretion of their

Civet cat

anal glands ('civet'), which looks like butter, has been used over the centuries as a base for perfumes. The reason is explained very succinctly in a serious chemical paper: 'it possesses an olfactory component recalling the smell of human scalp and pubic hair' (Ward & van Dorp, 1981). The secretion is extracted every 2 or 3 weeks from farmed civet cats; animals are restrained by pushing them with their nose into a cone-shaped cage. Then 2 or 3 grams of civet are collected from the pocket-shaped anal glands with a small wooden spatula.

Most of the civet used in the western world came from Ethiopia, which produced it in quantities of 1 kilogram packed inside a buffalo horn. As recently as 1953 almost 30 tonnes of civet were exported from that one country alone, but then a big decline set in, and in 1963 only just over 1 tonne was produced. Even that smaller amount was the annual production of over 20 000 civet cats. The reason for the decline was that the essential constituents of civet can now be synthesized and produced much more cheaply in the laboratory. We do not need live civet cats any more.

TARGETS FOR SPORT

Perfumes may ease the wheels of our social lives, but there are further and more bloodthirsty ways for us to entertain ourselves with carnivores.

'I am Assurbanipal, the King of the World, the King of Assyria! For my regal amusement I have caught the Desert King by his tail, and on the instructions of my helpers, I have split his head with the two-handed sword'. Thus goes the translation of an Assyrian lion hunt, somewhere between the ninth and seventh centuries BC. Hunting for sport goes back into history even further than this, with Pharaoh Amenhoteb III killing 102 fierce looking lions during the first 10 years of his reign, around 1400 BC (Guggisberg 1962).

To this day, downing a lion is still one of the ultimate tests of manhood. It is an enhancement of status whether for a Masai warrior with a simple sword and shield, or for a well-equipped western tourist in Africa with a high-velocity rifle. Some people make it easy for themselves, sitting up over a bait, others expose themselves to considerable danger by following the animals on foot. Even in the Serengeti National Park in the 1960s, visitors such as the Yugoslavian leader Tito were given permission to shoot lions, just for fun. Some hunters clocked up large numbers of lions – people such as John Stevenson-Hamilton who had a record of 200 to his name in the early part of the twentieth century.

Today such figures may sound horrifying, but they are still small beer compared with what happened to tigers in India. George Schaller documented the Maharaja of Surguja as having bagged 1150 tigers by 1965, the Maharaja of Udaipur shot over a thousand, and the Maharaj-kumar of Vijayanagaram was catching up with 323. The British King George V was a mere amateur compared with them, but even so, in 1911 he managed 39 tigers in 11 days of hunting in Nepal. The Maharajah of Nepal, in contrast, finished off 433 tigers between 1933 and 1940 (Schaller 1967). Mostly the quarry was shot from the back of an elephant or from a platform near a live bait, such as a cow or a goat tied to a tree. The procedure carried little risk for the hunter.

There is nothing utilitarian in such hunts, and usually there are no excuses such as harvesting for fur or destruction of cattle raiders or maneaters. The hunting is done for sport only, for fun, with trophies as proof of prowess. More recently, hunting has become something of an industry, one that employs people who organize safaris, hunter guides, and people in game management. It brings in money through permits, travel, and in many other ways. Travel and hunting agencies advertise shooting safaris for many countries including South Africa and Tanzania. Clients have to pay for numbers and species shot, with the lion and leopard being by far the most expensive. A hunting permit for these animals requires $3500 per head in South Africa and $2000 in Tanzania, on top of the safari costs, licences, etc. (1998 prices). Fortunately, tigers are now protected, at least officially.

In Europe and North America hunters pursue bears for sport. There are dozens of sites on the Internet advertising bear hunting with rifle or crossbow, with an experienced guide, ranging in cost from about $8000 to $12000 for a bear in Alaska to about $1200 further south. An Alaskan wolf or a wolverine sets you back no more than $4000. Bears are usually shot from a blind or a high seat over a bait.

In the technologically more simple days of the Roman Empire, of course, the free-choice element was a bit more evenly distributed between man and beast. Battles between lions and gladiators or slaves were organized in the arena, and the Bible tells us how the prophet Daniel was saved by his faith during just such a confrontation.

Shooting badgers for sport is a popular pastime in many European countries, although now outlawed in several. The animals are shot near their dens ('setts'), or they are dispatched after being chased and surrounded by dogs. Considerable numbers of animals are involved: for instance in 1989, Swedish hunters took almost 40000 badgers, Norwegians about 9000, in Switzerland the score was about 1500 and in Poland about 1000 (Griffiths 1993).

In Britain people still organize badger baitings. Badger digging and baiting, even as recently as 1997, allegedly claimed the lives of hundreds of badgers, despite their protected status (Brace 1998). It is difficult to get accurate figures, because the activity is run subversively. Many hundreds of people are involved, often near the larger towns, and in all parts of the country. It is totally illegal, but it still happens frequently, just as in previous centuries baiting bears with dogs was popular.

From what I can gather, badger baiting is usually carried out as follows. First, a party of diggers go out, often at night, with their spades and terrier dogs. Badger setts are easy to find, and they are often known to the men. A dog is sent down one of the entrances, and, once an animal is pinpointed underground in its sett by the terrier, the diggers break through the roof of the sett whilst the dog stops the badger. The animal is grabbed with a special pair of badger thongs and dumped in a sack. Often the badger is deliberately maimed by breaking its jaw or a leg, and then released either near its sett or in a special enclosure (often made of straw bales). Dogs are then freed to attack the badger, and bets are placed on which dog will deliver the fatal bites.

In Alaska, a quite different sport was the shooting of wolves from an aeroplane, when they are exposed in the open on snow or ice. After years of large-scale slaughter, the sport was regulated in 1963 and recently the rules allowed only two wolves to be shot per aircraft per year. It is likely that more restrictions are to come because there is much public opprobrium. As David Mech comments, 'To many hunters shooting wolves from an airplane is the ultimate in sport, but to others the activity is the lowest form of recreation imaginable' (Mech 1970).

An old and now also highly controversial sport is fox hunting in Britain and a few other countries, famously tainted forever by Oscar Wilde's description of 'the unspeakable in pursuit of the inedible'. This suggests that fox hunting is enjoyed only by the upper classes, but nowadays enthusiasts also come from other walks of life. A pack of hounds, followed by a 'field' of appropriately dressed followers, mounted on horseback (as well as people on foot and in cars), chases a fox or a succession of different foxes often over many miles. At the finish the fox is killed by the lead dog or torn apart by the pack, but usually it manages to escape, often by going to ground.

I will mention a few statistics to give an idea of the scale of fox hunting. In the year 1998, there were about 200 recognized 'fox hunts' in Britain, i.e. organizations with packs of hounds and a following on horseback. Some 215 000 people followed the hunts, and up to as many as 200 horses took part in any one hunt (more typically 50). It is estimated by anti-fox-hunting organizations that about 20 000 foxes are killed by hunts each year in Britain (although this figure appears very high to me), out of an estimated total fox mortality from all causes of 400 000 animals per year. In a sample of 421 'finds' (encounters of a hunt with a fox), 14% of the foxes were caught, 50% outran the hounds after an average chase of 31 minutes, and the rest escaped into a den, of these a few were then dug out with help of terriers and shot (Pye-Smith 1997). In terms of numbers, therefore, fox hunting does not make any impression on fox numbers, even by the possibly rather extreme estimates of those against fox hunting: it is just the odd few that get caught. Fox hunting is a sport, and not a management activity as is sometimes claimed by protagonists.

Although it may be a sport, fox hunting attracts a great deal of public protest because of the animal suffering involved, and it seems highly likely that within the next few years it will be banned in Britain by law. Even as I write this, there is a firm attempt by the government to abolish it. There also used to be otter hunts in Britain, but they were made illegal in the 1960s, because of concern about otter numbers in the country (although otters declined because of pesticides, not because of hunting). Former otter hounds are now trained to be used on mink, of which they kill some 600–700 per year (Pye-Smith 1997).

It is very difficult to obtain figures on the contribution of the hunting of carnivores for sport to the economy of countries. However, whether we approve of sport hunting or not, there is no doubt that it is economically important in terms of the income and jobs involved, and there is no question about the importance of carnivores as the objects of pursuit in many countries. Nevertheless, there appears to be an increasing sentiment against such use of carnivores. The animals may cause us trouble, they may threaten us, but for some reason most people do not want to see them killed, at least not for 'sport'.

CARNIVORES AND ECOTOURISM

Fortunately (in my eyes) cameras have replaced rifles in most forms of wildlife tourism. Ecotourism is booming, increasingly so. People visit wildlife reserves and national parks, and the standard advertisements feature visitors taking photographs from vehicles clustered around a cheetah or a pride of lions, people on a platform watching bears, or from the back of an elephant looking for tigers. The trophies are flat and rectangular in those cases, but no less real than the skins with gaping mouths in front of the fireplace. The income from wildlife tourism is huge, and many of the largest national parks have demonstrated that the most important reason why people want to come to the national parks is carnivores. More than anything else, visitors want to see lions in the Serengeti and Kruger, bears in Alaska, cheetahs in Amboseli, tigers in Kanha.

As an example, in a survey in 1994, visitors to the Kruger National Park in South Africa were asked specifically which animals they came for. The lion got more votes than any other of the scores of species there: the animals people most wanted to see, from the safety of their vehicles, were carnivores (52% of respondents), especially the lion (29%), but also the leopard (11%) and the cheetah (9%). Only then were others such as elephants mentioned (9%), and these can also be pretty dangerous; almost none of the many other ungulates or birds in the park even got a mention (Winnikof 1995). I will return to this attraction of carnivores for visitors in later chapters.

FUR TRADE

We have seen that, over the millennia, mankind has appreciated carnivores for their meat, for their bones and for their glands. We have enjoyed hunting them with spears, arrows, guns and cameras. But the one product of wild carnivores that has always been more important to us than anything else is their fur. It still is.

One cold, blustery day in February 1996, with a temperature well below −20 °C, I found myself hurrying along the pavements of Minsk, in Belarus. The streets were wide wintry deserts, with huge concrete structures towering above the many Belarusians. It was a city scene repeated many times over in towns in Eastern Europe, the people scurrying through slush, between huge heaps of dirty snow, wrapped up against the biting cold. Every single person wore a shapka, a hat, the traditional Russian garment, and from what I saw around me in town, more than half of these shapkas were made of real fur.

A hat is a life saver there, a necessity, and animal fur is still the most efficient material from which to make it. The best, most expensive furs are those of carnivores, and I noticed that the shapkas on the pavements of Minsk represented many of the East European species, such as mink, pine marten, sable, otter, raccoon dog, fox, wolf and lynx, as well as various other animals.

In the Belarusian capital fur coats are also fairly common (despite being expensive), and may be worn by both men and women. Unlike fur coats in many other countries, the material is used not just for its looks, but because it is functional. Fur is a commodity that is traded in large quantities, and fetches high prices. With the advent of synthetic materials it is perhaps no longer much of a necessity in rich, western countries, and it is certainly not really needed in the tropics. But in Eastern Europe it is clear why furs have such appeal, and even why people with a fur coat or shapka may well have a higher survival rate than people without. Further north, especially in Arctic regions, furs mean survival, and until recently the Inuit were totally dependent on them; seal skins, wolf skins, bear skins, Arctic fox skins, every species had its use. Wolverine skins, for instance, were in great demand because they are excellent for lining parka hoods, as it is easy to remove the frost from them.

The American fur trade asserts that about one in five women in the USA owns a fur coat (a claim that seems rather steep to me). However common fur coats may be, they are often worn just for looks rather than protection. They are obviously highly functional in extreme climates, but much more often, at least in the West, fur is just glamorous, with an important role as an expensive status symbol.

Several of these points are clearly emphasized by the many artificial products that emulate furs. Leopard skin patterns are ubiquitous,

Wolverine

and where would English pageantry be without the bear-skin hats of the Guards (except that they do not use bear fur any more)? Interestingly, the look of fur rather than just its insulating quality has been appreciated only since the nineteenth century. It was then that people started to wear animal skins consistently with the fur on the outside; this is much less effective in keeping out the cold than having the fur on the inside.

Worldwide, the scale of the fur trade is staggering. In 1995, fur retail in the European Union alone stood at $6 billion, employing 175 000 people directly, with an extra 50 000 in the supply trades to the fur industry. In the same year, the fur retail trade in the USA was $1.2 billion ($1.27 billion in 1997), employing 50 000 people. Canada has about 80 000 fur trappers, most of them indigenous people, contributing about $600 million to the economy (http://www.iftf.com). Carnivores produce most of this revenue (some others are beavers and muskrats).

I will just quote a few more random examples from earlier years. The 1977–78 winter harvest of carnivores in North America alone brought the trappers $166 million. A lynx skin in the 1980s fetched about $400 as sold by a Canadian trapper, and in some years more than 40 000 lynx were caught annually in Canada alone. This was a value of $16 million to the Canadian trappers, which increased to several times that amount once the skins were fashioned as fur coats and other garments (Funderbunk 1986). At the top-end of the market, in 1975 a single high-quality jaguar coat fetched $20 000 in a fur boutique in New York. In Tokyo in 1980 a clouded leopard coat was sold for an unbelievable $124 270, a leopard coat for $72 000 and a tiger coat for $94 700 (Eltringham 1984). In 1978 the export of grey fox skins from Argentina was worth $33 million, whereas in that year red fox skins from UK and Ireland traded for around £2 million ($3 million) (Macdonald & Carr 1981).

Some of these trade and export figures cannot claim to be strictly accurate, based as they are on various assumptions, and excluding some parts of the trade whilst guessing at others. But they do show the order of magnitude, as well as the large significance of the fur trade to economies everywhere in the world. When pressure groups campaign against the use of furs, they are up against a gigantic industry.

Most fur of wild animals originates in the USA, Canada, Russia and China. But many smaller countries are also involved, especially in South America and South-East Asia. The fur trade in those small countries may be relatively more important than amongst the big industrial giants, even though it may involve species for which trade is outlawed. To mention just one example: many spotted-cat fur coats are sold to this day to tourists in Nepal, despite the protected status of the (common) leopards, snow leopards, clouded leopards, lynx, leopard cat and other animals involved (Heinen & Leisure 1993). The tourists who buy these coats are mostly from Spain and Italy (hardly countries where the climate demands the use of furs), although the large majority of visitors to Nepal come from English-speaking countries.

During fieldwork in my recent study on European mink in Eastern Europe, I have come to realize that carnivores are trapped almost everywhere in eastern regions, however remote or urban. Skins are sold locally or through official government channels. Any carnivore goes: European and American mink, otter, fox, wolf, lynx, raccoon dog, with prices varying over the year. In 1996, a wild-caught mink skin fetched US$60 in Belarus, which was a large proportion of someone's monthly income. The revenue from trapping can be quite good, and for many people in the East European countryside, trapping is almost a necessity during times of economic stagnation or unrest. The trappers, there as everywhere else, are highly skilled, with lifetimes of field experience, and with almost religious beliefs in their methods and observations.

Many different species of carnivore are involved in the fur trade. The most comprehensive data are for 1978, and the following list shows the numbers and variety of skins of wild-caught animals traded in that year in the USA and Canada (Nilsson 1980; Macdonald & Carr 1981):

Raccoon	3 913 000
Coyote	445 000
Red fox	389 000
Mink	350 000
Grey fox	265 000
Striped skunk	175 000
Bobcat	87 000
Ringtail	76 000
Otter	47 000
Long-tailed weasel	46 000
Arctic fox	37 000
Short-tailed weasel	36 000
Badger	34 000
Lynx	23 000
Fisher	14 000
Spotted skunk	12 000
Wolf	7000

Given the high prices of fur, it is not surprising that illegal trade in skins flourishes, despite international agreements such as the Convention on International Trade in Endangered Species (CITES). For instance, in 1976–77 one dealer smuggled into New York the skins of 30 068 ocelots, 46 181 margays, 15 470 South American otters, 271 giant otters, 5644 leopards, 1867 cheetahs, 1939 jaguars and 468 pumas. All of them are protected and excluded from international trade under the CITES agreement, and collectively they are worth a fortune (Eltringham 1984).

Over and above the wild 'harvest' there are also the products of fur farms. In terms of quantities, farm furs are much more important than wild furs, and they contribute about 85% of the turnover in the fur industry. Most of this (64%) takes place in north-west Europe (especially in Norway and Denmark), the rest in North America (11%) and in Russia and South America (http://www.iftf.com). The American mink is the most common fur-farm animal, and worldwide production in the 1980s stood at almost 30 million pelts per year, fetching prices of $30–50 each. Some special and difficult breeds of mink have a much higher price tag: for example, the breed called 'Kojah' fetched $2700 for a single skin in the 1960s (Dunstone 1993). Apart from the mink farms, there are also many fox fur farms in several countries. A huge business is involved, with vast numbers of carnivores farmed. In Britain I visited a mink farm with over 13 000 animals, which was a small operation by international standards.

The fashion for furs changes over the years, fluctuating between trends towards long-haired furs, such as those of the raccoon, fox and wolf, and trends towards short-haired furs, such as those of the mink and otter. However, the last 30 years have seen a tendency in western countries to abandon altogether the trade in skins of popular species such as spotted cats and otters, for ethical reasons. Also, and for the same reasons, many people simply do not want to use any furs at all. There are demands to abolish trade in skins of wild-caught animals, and international pressure has built up to stop the trapping of animals with leg-hold traps. But ethical arguments are replaced by others, and recent newspaper articles report a renewed upsurge in the fur trade, after it went through a few years of uncertainty.

Carnivore hair is also used for other purposes, for instance in making different types of brushes. The French name for the badger is *blaireau*, the same as their word for shaving brush, and the connection is obvious. But that connection has also become obsolete now that the use of badger hair is no longer allowed, partly because badgers acquired protected status, and partly because of health reasons where badgers are one of the species susceptible to rabies. Painters say that no paint brush beats those made of hair of the sable, but they, too, are now largely superseded by brushes with synthetic hair. In Scotland people use the head skin of otter, badger or wildcat as a cover for their 'sporrans' – the pouches worn in front of the kilt. But in the end, fur for clothes is the only really important use of carnivore hides.

Things have moved a long way from the simple animal-skin garments used by our prehistoric ancestors. Nevertheless, we still see the same exploitation by man, involving largely the same species of carnivores. The strong movements against the fur trade on ethical grounds are fuelled by arguments against the principle of animal exploitation, or concerning the welfare of animals kept in small cages in fur farms, and of wild-caught animals caught in cruel traps. Interestingly, people object to keeping animals in fur farms, but there are almost no such concerns about animals reared for meat (and kept in equally bad conditions as exemplified by battery hens or battery pigs). There are valid concerns also about the survival of wild populations, especially of the spotted cats. Partly in response to all this, and partly for economic reasons, cheaper, artificial products are replacing natural furs.

Furs stir very strong emotions in people. Spectacular and very vocal campaigns are waged in western countries against the use of furs:

a good example was a recent television advert of a fashion show, with fur-wearing models leaving trails of blood on the cat walk. In western countries the wearer of a fur coat even invites activists to spray paint or throw dirt.

Personally, I would not wear a fur hat or coat, but I have tried to discuss the issues objectively. I think it is fair to conclude that, whether we approve or not, it seems likely that exploitation for the sake of the fur is an important element in the relationship between carnivores and man, and that, whatever their immediate importance to some of us, there is no doubt that furs will be here for a long time to come, whether originating from wild animals or from farms.

Much of this chapter shows that there are many different ways in which we benefit economically from wild carnivores. Hunting them for sport is big business, and so is the fur trade. But the importance of hunting these predators appears to be declining, and it is disapproved of by an increasing number of people, particularly in western society. In many countries furs are also less important now than they were formerly. The other uses of carnivores, such as for food or medicine, may be dramatic, but in western countries they are insignificant in economic terms, and there are strong pressures on other countries from western groups of conservationists and politicians, to put a halt to this exploitation.

More and more, the emphasis is on non-consumptive enjoyment of the carnivores, for instance through wildlife viewing live or on television, and through photography. Increasingly, we want to protect the animals. Rationally, we see the contributions of carnivores to the diversity of ecosystems as a reason for maintaining their presence. But for many people it is mostly the beauty of these animals that fuels reactions to them, and the appreciation of their character (through our acquaintance with pets), but not their use to us in economic terms. Nevertheless, the material importance of carnivores will always be a considerable factor to be reckoned with.

Uses for dogs

8

Wolves with human souls: pets
Dogs and cats as working animals and companions

In November 1997 the then President of the USA acquired a dog, just an ordinary dog, and it was the first time the man had ever had one. The animal made headlines all over the world, and the acquisition was seen rather cynically as a major attempt by Bill Clinton to improve his image amongst the populace. One remembered President Truman's observation that 'if you want a friend in Washington, get a dog'. Only a few presidential terms before Clinton, pictures in the media of Lyndon Johnson pulling the ears of his canine companion had contributed significantly to his subsequent loss in the election. George Washington owned 37 dogs, and almost every president since then has had at least one. Also, the British royal family is often pictured surrounded by an entourage of corgis. Obviously, image-makers are in no doubt that pets have a very significant influence on the perception people have of each other at all levels of our society, and Clinton was a late learner.

All this is based on people's deep love of pets. Of course, they are not to everybody's liking, but for most of us dogs or cats are almost as close to our hearts as are other people, and someone who is nasty to a dog is an enemy. We look them in the eyes, love the feel of their coat, and we play with them almost as with children. There is great warmth between pets and us, and a feeling of disbelief if ever one remembers that one is holding a direct descendant of a wolf or a wildcat.

Newspaper reports on statistics released by the British government for 1997 showed that this country housed 7.2 million domestic cats and 6.6 million dogs, the combined figures equalling almost one quarter of the number of people in the country. Similarly, the USA counts 54 million dog owners, comprising about 38% of all households. In other countries pets are just as popular. Almost all pets are cats and dogs, with numbers of gerbils, hamsters, rabbits, horses, budgerigars, fish and whatever else people keep totally insignificant in comparison. Everybody will be aware of pets on an almost daily basis, even people who do not actually own one. Even most non-owners still like pets, they like touching them or talking to them. Why, and why especially carnivores?

Going back into prehistory to look at the place of domestic carnivores in human society will not answer these questions completely, mostly because the role of these animals, and the reasons for keeping them, have changed so much over time. There is little doubt that in the dawn of evolution of agricultural society the majority of cats and dogs were kept for good utilitarian reasons. Cats caught mice, and dogs had a variety of tasks as large as the number of different breeds. But nowadays, in western society and in many other countries, the most common function of a pet is that of a companion, of a surrogate human being.

Both dogs and cats have been in service since long before we could write about them. Dogs were domesticated and bred from wolves. They were the first animal species to be domesticated by man, until recently we thought some 14 000 years ago, at the end of the last ice age when we were still wholly dependent on hunting and gathering, and long before livestock came into use. Knowledge about that first date came from deposits in a palaeolithic grave in Germany, where the fossils showed clearly that apart from people, there were domestic dogs, derived from the wolf but significantly different (Clutton-Brock 1995). Only several thousand years later (9000 BP, before present) did goats, sheep and other food animals follow dogs into the service of human society. However, more recent evidence from the DNA of domestic dogs suggests that in fact, domestic dogs were separated from the wolf very much earlier than was concluded from the fossil evidence, more than 100 000 years ago. This happened almost certainly by the hand of man, who started selective breeding (Vila *et al.* 1997).

Remains of dogs from all periods of prehistory since 14 000 BP have been found in many places, on all continents. The earliest dingoes were taken into Australia some 3450 years BP, by seafarers from South-East Asia (Corbett 1995). Since those early days, dogs have been consistently popular all over the world, and their images are found in paintings and writing from every period of history.

The domestication of cats came considerably later than that of dogs and other animals, with the earliest clear evidence (a painting) from around 5000 BP, in the fifth dynasty of pharaonic times in Egypt. The Egyptian house cat was almost certainly derived from the local wildcat there, *Felis silvestris lybica* (Serpell 1988). This is a subspecies of wildcat that also nowadays is relatively easy to domesticate, and in Africa people still catch wildcats every so often and train them successfully. In the deserts of northern Kenya I saw people with domestic cats looking exactly like the wild ones there, and I was told that often wild kittens were brought into domestic service. The Egyptians venerated the cat, and they tried hard to prevent its export to other cultures, even employing special squads to retrieve some illegal exports. Greeks and Romans initially used ferrets and polecats for catching mice, in preference to cats, but before long there was no stopping the use of cats and they spread to everywhere around the globe.

Cats are basically a solitary species, moving silently at night rather than by day (unlike dogs). Even domesticated cats largely go their own way. Such characteristics were the reason for their association with witchcraft in the Middle Ages, and for about four centuries public opinion in Europe moved strongly against cats. Black cats in particular became the victims of persecution, and James Serpell's study on their history mentions that 'on Christian feast days all over Europe, as a symbolic means of driving out the Devil, they were captured and tortured,... all in an atmosphere of festive merriment' (Serpell 1988).

Thank heaven those days have gone; in western society people have lost much of their fear of the dark and of witchcraft, and during our time cats continue to be popular as companions. They have been largely dissociated from evil, and generally they are back on the pedestal where the Egyptians once put them. However, black cats are still considered a bad omen in many countries, although for some perverse reason they are now a lucky sign in Britain.

History, therefore, shows that both dogs and cats have had a long period of coexistence with man, embracing hundreds of human generations. The association has lasted long enough to affect and mould people's image of all carnivores, including the wild ones.

TRAINED TO WORK

Very early on in the history of domestication selective breeding must have occurred, as the first specimens of dogs in human burials were recognizably different from wolves. Nowadays, moving through any town or village, anywhere in the world, one meets a bewildering variety of cats and dogs in a multitude of sizes and shapes, a variation on a par with that within the whole order of carnivores, although as domestic animals only a couple of species are involved. The range of sizes, shapes, colours and hair lengths is especially large for the dog, and I still think it amazing that they all have been bred from the magnificent wolf. It is difficult to comprehend that a Pekinese, a bulldog and an Afghan are derived from one and the same species, especially after one has been exposed to arguments from taxonomists showing that very similar looking wild jackals in Africa are three totally different species.

The dog's parent species is probably inherently very variable: at the extremes, there are black and white wolves and small and large wolves, sometimes even within one pack. Furthermore, wolves have probably been domesticated on separate occasions in many different parts of the world, from India to Europe and North America. Domestic cats, in contrast, are traced back to a much smaller area, and their wild forebears have far less genetic variation in external appearance to build on.

It must be partly owing to this large inherent genetic variability in shape and character of domesticated wolves that our ancestors have been able to breed the species into products which can be used for an almost endless range of different purposes. Cats had and have fewer utilitarian functions in human society, catching rodents being the most important. One of their most exotic roles must be catching vampire bats: they are used for this in Argentina when the bats feed on the blood of livestock and people (Delpietro *et al.* 1994).

The variety of duties that dogs have been taught and for which special breeds have been created is mind-boggling. Hunters use dogs for finding, chasing and stopping a quarry so that it can be dispatched by their much slower selves. Pointer dogs and retrievers help hunters in a different way, as their name suggests. Terriers go to ground to fetch subterranean prey or pin it down for capture by man. Other breeds put up wild birds and mammals so that they can be shot in flight, whereas greyhounds and lurchers catch fleet-footed prey (such as hares, gazelles or roe deer) for their owners.

In many countries sheep farming would be quite impossible without dogs. They have two different roles in this industry, depending on the country. Firstly, where there are no wolves (e.g. in Britain or Australia), sheep are scattered over large areas and left to roam, and collie dogs are used to gather the sheep off the hill or help farmers move them around. There is an amazingly complex interaction between dog, dog handler and sheep. A shepherd handles one or more dogs by means of just a few commands, whistled or shouted, and between them man and dog move flocks of sheep between fields and through gates. The dogs run and drive the sheep over long distances, circling and occasionally snapping at them, they are fast and highly efficient, and no person or vehicle could ever emulate them. The ability of a dog only shows after many months of training, but it is also genetically determined, and sheepdog puppies are selected for training from litters of 'good' parents, each dog worth a fortune.

Secondly, in countries where wolves or coyotes are present, large dogs are trained to protect both sheep and cattle. Breeds such as the German Shepherd and Abruzzo, Maremma, English, Pyrenean and Anatolian sheepdogs stay with the sheep or cattle at all times of day, and their charges graze close together in dense, well-guarded herds. The dogs themselves are often protected against wolf attack by ferociously spiked collars.

In the semi-deserts of northern Kenya in the 1970s, the pastoralist people in some of the nomadic manyattas suffered badly when temporarily they had no dogs. Their animals had been culled by the government's veterinarians in a campaign to control rabies, and every day these settlements sustained damage to their goats, sheep and cattle herds from hyaenas, jackals and lions. But in each of the other manyattas where I was met by barking, snarling and biting dogs, the people reported that they had had few problems with wild predators, and the difference in predator damage between manyattas with and without dogs was highly significant (Kruuk 1980). In North America in 1980 a project with sheep farmers in Idaho concluded that the initial cost of acquiring and training a guard dog (about $900, plus 600 man-hours) was really worth it: the expenses were more than offset by subsequent savings of about 70 sheep and goats per farm per year – animals that would otherwise have been taken by coyotes, feral dogs, bears, cougars, wolves and others (Green & Woodruff 1984).

There are guide dogs for blind people, police dogs to track or subdue criminals, guard dogs for security services or personal protection, greyhounds earning thousands of pounds or dollars in dog races, search dogs to find truffles, St Bernard dogs to rescue people from snow accidents in the mountains, huskies to provide vital transport under arctic conditions or for entertainment in sledge races, the dog that pulled the milk cart when I was a boy in Holland, the animals that sniff out drugs for the customs officer, and poodles that perform their dazzling tricks in a circus. One can go on and on through an array of animal duties almost as wide as the human imagination.

DOGS AND CATS AS COMPANIONS

Despite all these practical uses, what is these days vastly more important than any of those tasks (at least in western urban society) is the role of cats and dogs in providing company to people. There is little doubt that originally this only played a secondary role, but now the *raison d'être* for the large majority of pets in many developed countries is companionship. An enormous industry has built up around this of which I cannot even guess the total worth, but it is many, many billions of pounds or dollars annually worldwide. It includes the trade in pets, pet food and equipment, veterinary care, shows, grooming, literature and media interest, kennels, graveyards, insurance, and a huge variety of other aspects.

When I ask someone why she or he has a pet, I often get some practical excuse. Protection is one of them, to stop burglaries or personal assaults. But clearly just the love of an animal is the main reason. Looking at this as an animal behaviourist, many of the arguments for having pets are based on the fact that we ourselves are group animals, and we create groups around us, of people or of pets if need be. Owners say they need something to love, they feel more relaxed when they can touch their animal (and there are no people around to oblige). Dogs provide play, they give and accept love, they provide emotional security, and they serve as child substitutes. People talk to their dogs or cats, they ask them questions, and a wagging tail or a purr can be taken as an affirmative or sign of agreement. Dogs also play a large role in facilitating contact with other people: owners go out more, and are more likely to talk to other people with dogs.

One detailed study concluded that pet owners generally have a higher morale, and they enjoy life more than do their pet-less peers (Hart 1995). It has been shown that petting a dog significantly lowers an owner's blood pressure (Jenkins 1986), and in institutes caring for people with Alzheimer's disease, patients with access to pets showed far less aggression and anxiety than others (Fritz *et al.* 1995). Children integrate demonstrably better into society if they are used to pets.

For several such reasons, pets have often become associated with status, and it is far from rare for people to produce excuses for *not* having a pet. One likes to be seen as successful, happy and well balanced, in control of one's immediate environment, hence one likes to be known as having a cat or a dog. Somehow, it makes a person's life more complete.

Although keeping cats is more common than keeping dogs, owners are more preoccupied and emotionally involved with dogs than with cats. For instance, researchers found that people spend more time interacting with dogs than with cats, they order them about more often, and there is more physical contact. In fact, almost half of people in pet-owning families have more physical contact with the dog than with other people. Pet owners usually think themselves dominant over dogs, but far less so over cats (Hart 1995).

Almost all of us anthropomorphize cats and especially dogs: we accredit them with many human attributes. It is not surprising, seeing how they can be taught, loved and dominated by us and fitted into our communities. People take pets to bed with them, and integration in

Pets

some societies, for instance in Polynesia, even goes so far that young puppies may be breastfed by mothers with small babies (Hart 1995).

What is so special about cats and dogs that they have been made into ersatz people? Perhaps most importantly, as companions they are often more tractable than other people would be. They are highly trainable, they do not dominate, they accept an owners' personality without comment or criticism, they can be commanded or cuddled, and they actively seek to synchronize their behaviour with ours (Bryant 1990; Hart 1995). We may pay a great deal of attention to them, but dogs are almost invariably even more attentive to their owners than vice versa (Hart 1995).

WHY CARNIVORES, WHY DOGS AND CATS?

Is there a reason why carnivores should be more suitable as companions than are other animals, such as non-human primates, ungulates or birds? Why did people not train and breed monkeys, goats or crows as companion pets? Many of these animals have an 'intelligence' that is at least comparable to that of carnivores, and they can learn a lot and fast.

Monkeys and corvid birds would be trainable, perhaps even more so than dogs, certainly more so than cats. Most ungulates would be far more difficult in that respect, even horses. All these other species lack one or more of the essential companion pet characteristics: to be not too big and not too small, intelligent and trainable, relaxed and not 'nervous' or too active, soft on the touch, sociable and ready for a cuddle. Candidates also should be fast breeders, to allow us to select suitable strains. Strong aggression is a characteristic that can make an animal unsuitable as a pet, but this can be bred out of a strain. Appearance probably does not come into it: I have always defended the idea that, at least amongst dogs, the ugliest animals are also the most-loved ones.

Carnivores have one other characteristic that suits us particularly well, in combination with their intelligence, trainability and adaptability: it is their activity, or rather lack thereof. Most wild carnivores, including wolves and cats, are normally active for only some 4 to 6 hours each day (and they are very happy with less than that). The rest of the time they sleep, although always ready for action if the need presents itself. Monkeys or crows are active throughout the daylight hours, always feeding, exploring and playing, at a level of activity that might drive us up the wall if we had to live with it day in and day out. The carnivore's sloth suits us much better.

One reason that dogs and cats were the carnivores chosen as universal pets was that they were already working for us and had been trained and domesticated, and additional companion roles were easily acquired. We have had, and still have, ample opportunity to experiment with other species, but to have familiar working animals at hand that already have had most difficult behaviour traits bred out of them must have been an overwhelming argument in their favour.

Moreover, wolves are pack animals, inherently inclined to synchronize and cooperate closely with conspecifics, with behavioural mechanisms of dominance and subordinacy built in. They, and their derivatives the domestic dogs, may be more suitable for their role as companion animals than the naturally solitary cats. But even the unsocial cats had, as working animals, been bred for useful traits as domestic companions.

Whatever the history, both dogs and cats are astoundingly successful as pets, and they clearly still are the most important animal 'tools' we have. Of course, several other carnivores also had, and still have, their turn as working animals. Trained cheetahs were kept for hunting gazelles in Persia and India, used in a similar fashion as were falcons for hunting birds: released near their quarry, they ran it down and killed it, then allowed their masters to take over. Ferrets, which have been bred and domesticated from polecats in prehistoric days and at least since early pharaonic times in Egypt, were used originally to catch rodents around houses. Now they are employed to catch rabbits: when the ferret is released into a rabbit burrow the rabbits respond by a quick exit, and they are caught in nets over the various tunnels from the warren, or shot when running away. In Bangladesh and China tame smooth otters are trained to fish, being released from a boat whilst attached to a leash from their harness, to chase fish into nets (Foster-Turley 1998). Several species of mongoose, civet or genet are kept for catching rats and mice in various Asian countries. What all these species have in common around the globe is that they are exploited for their efficient hunting behaviour, trained and adapted for use by people.

The explanation, therefore, of why most of our pets are carnivores, probably has its origin in hunting habits. It is one of the behaviour categories that man and carnivores have in common, and the main reason why we admire them. Our reasons for choosing dogs (wolves) and cats out of all other carnivores could be that natural prey size, hunting and social behaviour make wolves a more suitable hunting companion than almost any other, whereas the alternatives to cats (for catching mice) are mostly too small to make them easily manageable around our homes. There is also the minor point that many other carnivores such as ferrets or wild dogs are quite smelly, although that would probably not have deterred our ancestors.

PETS AND CONSERVATION

Pets have a bad name amongst conservationists. Not all of the reasons for this are relevant ones, but some of them express valid concerns. Letters to the editor in newspapers often scream outrage over the multitude of birds that are murdered in gardens by cats, and around villages and towns all manner of wildlife may be chased and killed by dogs. I have some sympathy with such complaints, but not too much, as in any case most of the victims are there because they are attracted to our fleshpots.

Much more serious is the threat posed by feral dogs and cats, especially on islands, and this I will discuss in more detail in Chapter 9. Even more of a stain on the reputation of domestic animals is the role of pets in communicating disease to wild populations, and one has become aware of this especially in recent years, with civilization encroaching on what was once wilderness. Studies in the Serengeti showed that canine distemper, a morbillivirus endemic in domestic dogs around the national park, caused fatal epidemics in lions, leopards, wild dogs, bat-eared foxes, jackals and hyaenas (Roelke-Parker *et al.* 1996). Similarly in the same area, rabies persists in the populations of domestic dogs around the (unfenced) national park, and these animals have frequent contact with wildlife (Cleaveland & Dye 1995). There was strong evidence that this domestic dog rabies was responsible for rabies outbreaks in wild dogs and other carnivores, and it was considered a very serious threat, especially to the survival of wild dogs (which since then have disappeared from the area) (Woodroffe *et al.* 1997).

A very similar situation occurs in the highlands of Ethiopia, where rabies is the most likely cause of the dramatic decline of Ethiopian wolves, the most threatened carnivores in Africa. Here again the cause is transmission of rabies from the fast increasing numbers of domestic dogs in the area (Sillero-Zubiri & Macdonald 1997). These Ethiopian wolves, incidentally, are also exposed to another, lesser danger from domestic dogs: hybridization.

Yet the role of domestic animals in wildlife conservation is not all negative. I think it is more than likely that when taking into account all aspects of keeping cats and dogs, including our emotional involvement and even dependence on them, that the integration of pets into our lives strongly affects the way in which we look at wild animals. We see the immediate and close wild relatives of our best friends through spectacles that have been coloured by experience with animals in our own homes.

Through such close contact with pets, people become aware of the differences between individual animals, aware of their 'personalities' and their charm, beauty and fascinating behaviour. This is likely to rub off especially on our views of wild carnivores, the animals that have so many things in common with our cats and dogs and look so similar. It comes more naturally to us to take a grand liking to wolves or foxes if we are accustomed to wrapping our arms around a big shaggy dog, and in the long run, loving one's cat may arouse concern about the conservation of its relatives.

Everybody can see individuality in our clever domestic companions. And there is no reason to assume that individual character differences are any less in wild animals. They are only more difficult to observe, and the animals look more alike. People privileged to be able to watch carnivores closely in the wild know that two individual wild hyaenas, foxes or otters or whatever, are as different as two domestic cats or dogs. We often recognize our wild study animals by their behavioural quirks, by their personalities. This individuality, I think, contributes enormously to the fascination these animals exert over us.

Conservationists have barely touched pet appeal and the personality of carnivores in their campaigns to involve the public. This is a shame, because our obsession with animals as expressed through the involvement with pets is firmly established. If carefully managed, this obsession may spin off considerable benefits to the survival of the wild brethren of pets, for it is surely easier to support the survival of a wild species if one can think of the animals as individual characters with their own habits and loves. Wild animals are then not just statistics, but become 'personalities' just as are our companions at home. On top of that, many of us would say that wild animals are that much more beautiful and attractive.

Lionesses and gazelles

9

Carnivores and neighbours: effects on prey

Effects on other species, and introduced exotics

It is almost impossible to evaluate all costs and benefits that came our way from the Carnivora. The animals give us great pleasure, warmth and companionship, we have eaten and used them, they have eaten and used us, and we have had to take costly measures for protection. But the chances are that on balance, over the period of existence of our species, we have lost to them, and that in economic terms they have cost us more than they have paid us back. We will see later in Chapter 11 that similarly, the carnivores themselves, in all probability, have lost more than they have gained from us.

It is useful, then, to see how other animals fare in the presence of predators. After all, just about every species suffers from predation. How, in general, do carnivores affect the numbers of their prey species? Often, one glibly talks about a 'balance of nature', suggesting that negative forces that act on populations are corrected by others. But what exactly happens, how does it work? Is there, indeed, a balance of forces or are appearances deceptive? These are important ecological questions in themselves, and they have to be answered, before in the following chapter, we can evaluate the anti-predator systems of animals in general, and the anti-predator behaviour of mankind.

'BALANCE' BETWEEN PREDATORS AND PREY

On the breathtakingly beautiful Serengeti plains in East Africa where, as George Schaller once put it, through the legs of an ostrich one can see mirages shimmering on the horizon, and where the clouds pile up above the huge diversity of literally millions of animals, it seems impossible that predators would really have any substantial effect on numbers of prey. The sheer natural opulence of the tremendous herds of ungulates grazing amidst the odd rocky outcrops and the flowers appears to defy the few lions, hyaenas or jackals in its midst.

The entire scene suggests that for the grazing animals, food is the only concern, as for cattle in a pasture. There is the appearance of peace, with the odd carnivore taking its dues, the unfit being replaced, and time has strengthened this impression. The Serengeti landscape was first engraved on my memory in 1964 when I started living there, and when I came again in 1995 after a 20-year absence it was all just as before. Charles Darwin could have been talking about the Serengeti when he said 'the forces are so nicely balanced that the face of nature remains uniform for long periods of time...' (Darwin 1859).

Yet such scenes are deceptive. There is a balance, there is an almost permanence about such animal communities. We see the stability, the lack of change. But we know now from bitter experience in many places that if even a small something changes, for instance if a new species invades, if man introduces a new predator, then everything may go haywire. Heaven knows what would happen were we to introduce a few packs of wolves or dingoes into the Serengeti. They might exterminate the entire population of one or more species in the area (alternatively, they might be slaughtered themselves).

Observations of such kinds of introductions with their bloody consequences, whether accidental or on purpose, have been made repeatedly over the twentieth century (see 'Artificial immigration' below). It has happened so frequently, that one conclusion must be that the balance of nature, the complicated numerical and long-lasting interaction between populations of predator and prey, is easily upset, at least by human activities.

The effects of introduced predators may be highly 'unnatural', but they are also significant for our understanding of natural events. It is probable that in their evolutionary history new species did move into apparently established ecosystems (such as the Serengeti), without interference by people, just in the natural course of events. From what we know of artificial introductions it is likely that new arrivals are usually snuffed out rather quickly. Maybe this also happened with past invasions without the help of mankind. Nevertheless, one cannot imagine that any ecosystem would escape successful immigration at some time or other.

Occasionally, we may witness a natural immigration event, usually unsuccessful. For instance, the South African carnivore ecologist and naturalist Gus Mills, who lived in the southern Kalahari desert, sometimes saw an impala there, hundreds of kilometres from its normal range of occurrence in the woodlands, and he met the odd baboon that had gone a similar distance astray from its usual habitat. Such animals looked healthy enough and there was ample food for them, but without exception the Kalahari carnivores took them out very quickly, actively selecting them from the large array of their regular prey species (Dr M. Mills, pers. comm.). There are other, necessarily anecdotal observations of such events. The new would-be settlers usually had no chance whatsoever.

Before such apparently 'stable' ecosystems were firmly established, there must have been a fair amount of to-ing and fro-ing of species, of potential prey or potential predators, and as I will show below, any immigration can lead to local extinctions. What we find today in any one place, therefore, is likely to be a remnant relationship only, the one ecosystem that is left after having been modified, with previous immigrants and populations wiped out by predators or competitors.

The observations of immigrating individuals (and of deliberate introductions, see below) suggest that many predator–prey relationships have only a small chance to get off the ground as a more equal interaction between populations. Any odd dispersing individual of a potential prey species is likely to be hit by predation long before a new prey population can arise, and most of such extinctions go unobserved. Predators can have a huge effect in this way, but it is subtle, and in the natural course of events almost undetectable.

PREDATION IN COMMUNITIES

Despite the negative effects of predation, rich assemblages of herbivore and carnivore species exist. In such communities of predators and prey one observes effective anti-predator behaviour, and a complicated relationship in numbers between carnivores and herbivores. Therefore, yes, there are balances, and probably numbers are regulated somehow, but this is relevant only to those species which happened to have survived previous extinction episodes.

I live amongst the foothills of the Scottish Highlands, and in the area immediately around our house there are foxes, badgers, wildcats, otters, American mink, stoats and weasels. It is a good variety, and between them they feed on most of the other vertebrates around, on roe and red deer, rabbits, voles, frogs, birds and others. Over the years changes in numbers of predators and prey have not been very conspicuous, just as in earlier days in the Serengeti, when we had about

25 different species of carnivore in just one area around the house, and about the same number of ungulate species for them to feed on, as well as many rodents and insects. Certainly, here in Scotland also, we see that there are fluctuations, and there is the odd disappearance and new arrival, but, in general, these communities keep going more or less as they are (if people keep their hands off them).

If one tries to understand why we do not get a long litany of disasters, of local extinctions and population explosions rather than mere smallish fluctuations in these non-island communities, several important mechanisms appear to play a role. Somehow in these stable ecosystems, predators kill without exterminating, and prey populations reproduce without numbers going through the roof. There is a large body of scientific research on the subject: many studies of the effects of predators on their prey species and vice versa. However, there are still big gaps in our knowledge.

At first sight, case histories of the effects of predation seem to originate from two different camps, although in practice the scenario is more complicated. One of the schools celebrates the 'doomed surplus hypothesis', and the other one points an accusing finger at carnivores, in the 'predator limitation hypothesis'. The 'doomed surplus hypothesis' goes back to American research in the 1940s. In a classic study (Errington 1946), Paul Errington demonstrated that in a population of muskrats preyed upon by mink it was mostly those animals unable to find the sanctuary of appropriate lodgings that fell victim to the predators. The result was a fairly stable ceiling to the numbers of muskrats, and a constant presence of mink.

A similar result came from the work of the Scottish naturalist Adam Watson and his colleagues (Jenkins et al. 1964). They showed that, from populations of red grouse on the heather moorland of the Scottish Highlands, it was mostly those birds that had not acquired their own territory that were removed by foxes, wildcats, peregrine falcons, harriers and other predators. The density of territorial birds was determined by entirely different factors, such as the quality of their food plants, and the number of grouse that were 'surplus' to that were removed by predation.

In general, however, this picture appears to be rather rare. There is now substantial evidence that prey populations may be strongly depressed by carnivores – evidence from removal experiments and also from 'natural experiments'. For instance, in Chapter 5 I have already mentioned an elegant project that was repeated over several years. On farmland in southern England, Stephen Tapper and his colleagues from the Game Conservancy removed virtually all foxes and crows from the study areas (Tapper *et al.* 1996). This more than doubled the partridge population at the experimental sites, compared with that in the control areas. It demonstrated beyond doubt the substantial depression of partridge numbers by these predators.

An outbreak of sarcoptic mange amongst Swedish foxes in the late 1970s and 1980s enabled the ecologist Eric Lindström to study the almost complete absence of foxes and their subsequent recovery, and the effects of these foxes on numbers of voles and different species of hares and grouse (Lindström *et al.* 1994). There were dramatic increases in the prey species, on a national scale and in detailed assessments on small study areas. These increases were reversed when the mange disappeared and fox numbers shot up again. Interestingly, this study also showed that, during the times when fox numbers and predation were high, the regular 4-year cycle in vole numbers had a large effect on hares and grouse. This happened because foxes switched to larger prey when fewer of their staple prey, voles, were available. However, when fox numbers were down because of the mange, the 4-year vole cycle had no effect on the other prey species. Clearly, the presence of the fox as the main carnivore puts a large stamp on the community of small herbivores in Scandinavia.

In many countries rabbits are an important agricultural pest, and there has been considerable interest in the effect of predators on their numbers. Could foxes or cats keep rabbit numbers down? In a review of all available evidence by the British Ministry of Agriculture, the answer to this question was a qualified 'yes' (Trout & Tittensor 1989). When rabbit populations were low, e.g. 1–3 per hectare, predators were able to maintain numbers at that level for years. In an experiment in Australia after a severe drought, rabbit density stayed where it was in two separate study areas of over 7000 hectares, each with foxes, cats and dingoes taking their toll of the bunnies. But on two other experimental sites predators were eliminated, and rabbit numbers multiplied by a factor 4.5 and 1.9, respectively, within a few months (Newsome *et al.* 1989).

Similar observations were made in New Zealand and in Britain. When rabbit numbers are at a high point (and that may be at 20–30 animals per hectare), predation by foxes, cats and others appears to have no effect whatsoever. Thus, the evidence suggests that predators can limit rabbit numbers if they operate in conjunction with other factors such as a disease or rabbit control operations. The overall effect of all this is that when one compares areas with and without intensive predator control, rabbit numbers are higher on land with few or

Leopard

no predators. Because of this, rabbit numbers are especially high on English and Welsh islands without predators, much higher than on the mainland. Despite their role in keeping pests at bay, carnivores are usually persecuted on the mainland, and the absence of predators there also means that there are more rabbits eating away at crops. There is little doubt that if one wants to keep fox numbers down, then crop damage from rabbits may be a price one has to pay for this.

Large predators in Africa are also likely to have substantial effects on prey communities, effects which are often more subtle than we would expect. In a stable grassland ecosystem of non-migratory, grazing animals (e.g. in the Ngorongoro Crater, in Tanzania), spotted hyaenas probably keep populations of wildebeest and zebra below numbers that would overgraze their pastures. Predators increase their predation rates when the physical condition of the ungulates deteriorates: they kill more and leave more for scavengers.

In contrast to these observations in the Ngorongoro Crater, hyaenas on the neighbouring Serengeti plains cannot maintain such pressure on the migratory population of herbivores there, because the predators are just not mobile enough despite their habit of 'commuting' over long distances to the big herds. Hyaenas have to return to their cubs in dens, since the young cannot move with the herds (Kruuk 1972a). For the same reason lions have little effect on the numbers of Serengeti wildebeest, zebra and gazelles (Sinclair 1979; Mduma *et al.* 1999). Despite its reputation, the Serengeti has relatively few predators for the huge herds of grazers and, in consequence, Serengeti ungulates often die of starvation, unlike those in Ngorongoro.

The non-migratory, resident populations of wildebeest and zebra in South Africa's Kruger National Park are hit much harder by carnivores than the shifting herds in the Serengeti, and Gus Mills and his colleagues there showed that lions probably limited the numbers of these ungulates (Mills & Shenk 1992). In North American ecosystems with resident ungulates such as deer, moose and caribou, similar roles have long been suggested for wolves and coyotes. They keep prey numbers below densities that would cause habitat deterioration (Longhurst *et al.* 1952).

But numbers are not everything in a community, and at least in the African savannah ecosystems predator effects can be profound even if they do not change prey population density, for instance in the Serengeti. There, many different species of ungulates have a lot in common in their choice of food plants. One might expect them to avoid each other in order to get maximal benefit from their grazing. Not so: the well-known ecologist Tony Sinclair, who has studied the Serengeti ecosystem longer than anyone else, showed that many grazers actually seek each other out and prefer company, finding safety in numbers. This happens especially in those places where predation by lions, hyaenas or leopards is likely. All the local movements of the grazing animals, of the wildebeest, zebra, gazelle, impala, hartebeest, topi and others, are demonstrably affected by the presence of predators (Sinclair 1985). This will affect their grazing, and therefore their food intake and their survival. The effects of predators reach much further than what we see in direct mortality.

Predator effects may be particularly harsh when they concern a favourite prey which is present in low numbers. For instance, in north-western Europe badgers eat mostly invertebrates, especially earthworms, and badger numbers are dependent on earthworm density. Worms are their staple diet. However, badgers are also extremely partial to hedgehogs. The result is that in England there are almost no hedgehogs wherever badgers are common, but there are many hedgehogs where the predators are absent. An Oxford research group showed through careful translocations and radio-tracking that predation by badgers was responsible for the lack of hedgehogs in many woods around Oxford (Doncaster 1992). Badger numbers were limited by earthworms, and hedgehog numbers by badgers.

In a comparable relationship in Australia, foxes appeared to be able to exercise a huge effect on the marsupial fauna: fox numbers were sky-high because they were dependent on the almost limitless number of rabbits (Saunders *et al.* 1995). Similarly in western Canada, wolves that depend largely on moose for their survival and whose numbers are limited by moose, are driving several (low density) woodland caribou populations to the brink of extinction (Seip 1992). It is observations like these that make me wonder what numbers of other herbivore species would have occurred on the Serengeti plains (and elsewhere), were it not for all the hyaenas and lions and jackals and other sharp toothed animals.

There is little doubt, therefore, that depression of prey populations through predation by carnivores is common. Tony Sinclair calls it 'a trite observation that in almost every case predation acts as a limiting factor' (Sinclair & Pech 1996). The question in which he is really interested is whether predators actually regulate populations of mammals, as distinct from merely depressing them. Do predators keep prey numbers within given bounds by increasing prey mortality when density gets too high, and decreasing it when numbers are too low?

As Sinclair points out, our knowledge here is sadly lacking: 'Large mammal populations may well be regulated by predators, but this has not yet been demonstrated.' However, there is no such doubt about other ecological factors in the life of these animals. There is ample evidence, for instance, that food resources can play a major role in actually regulating the numbers of almost all species, causing numbers to go up as well as down (Mduma *et al.* 1999). Whether diseases can have similar importance is still quite unknown.

LIMITED NUMBERS OF CARNIVORES

So, on the one hand carnivores may have far-reaching effects on prey populations, but on the other these predators themselves are often limited by their resources, often by food. The effect of the availability of prey on predators has been the subject of many studies. One of the first researchers in this field, the father of ecology Charles Elton, found that fluctuations in the numbers of lynxes trapped in Canada closely trailed the variation in numbers of their main prey, snowshoe hares, which showed a cycle of 10 years' length (Elton & Nicholson 1942). Many other studies followed, and the dependence of animal populations on

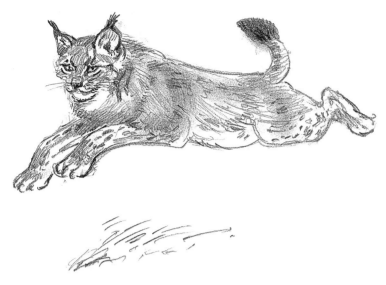

Lynx

their food supply was formalized by David Lack in his classic *The Natural Regulation of Animal Numbers* (Lack 1954). This drew mostly on results from studies on birds, but the ornithological conclusions were very relevant also for carnivores.

A few further examples show the general picture. In Minnesota, USA, wolf density was closely correlated with numbers of white-tailed deer (Mech 1970), and in Wyoming the sharply declining populations of black-footed ferrets followed the demise of prairie dogs (Clark 1989). Two researchers in Scotland, Ray Hewson and Hugh Kolb, studied the fluctuations in numbers of foxes there, in order to establish the effectiveness of fox control for sheep farmers. They found that the numbers of foxes killed closely trailed the numbers of voles caught in their traps, in a clear 4-year 'vole cycle': when vole numbers were high, numbers of foxes killed peaked, and when voles were few, so were foxes (Kolb & Hewson 1980; Hewson 1981). The effect of 'fox control' was almost negligible.

In East and South Africa spotted hyaena and lion numbers depended on numbers of wildebeest (Kruuk 1972a; Mills 1990; Mills & Shenk 1992), and in China George Schaller and his colleagues found that giant panda numbers were related to the abundance of bamboo (Schaller *et al.* 1985). In Scotland I estimated denstities of otters and fish and found them to be clearly related (Kruuk 1995); there are scores of other, similar results.

However, it is only rarely that our knowledge about the relationship between numbers of predators and prey is really sound; only rarely do we have more than just diet information, and a good correlation between fluctuating numbers of predators and prey. Unfortunately, there is almost no opportunity to test a relationship experimentally with wild carnivores. Nevertheless, the results of many case studies leave little doubt that food (i.e. prey) is frequently a limit to numbers of predators, though at times this effect is obscured by other habitat requirements or mortality factors.

In some species we have an idea about how this prey effect on carnivores works, and what happens when food is short. When predators catch prey, it may be a large, energy-rich prize. But the effort involved in obtaining it may be huge: a long, fast and exhausting chase, or an hours-long and highly tense stalk, a deep dig, or a long dive in ice-cold water. It has been possible to estimate the effort for otters and for African wild dogs, by measuring their metabolism whilst hunting, and to compare this with their energy gain, i.e. the prey caught. The results showed that such predators often play for very high stakes, a high-risk game: they may gain an energy-rich prey, but because their hunting is such a demanding exercise, the risk is a poverty trap.

If, for instance, a decline in available prey occurs (e.g. fish for an otter), then immediately the predator has to work much harder, and it has to spend much more extra energy to acquire prey. In consequence, it needs more food to sustain itself and meet the increased demand, therefore it has to work harder still, and so on. When hunting is costly in terms of energy, as it is for these particular animals, then the amount of time which a predator needs to forage each day curves up very steeply in the face of a declining food supply (Kruuk 1995; Gorman *et al.* 1998). This is a problem which is much more acute for a hard-working wild dog or an otter than it is for a browsing deer. Starvation is the penalty.

The effects of carnivores on prey populations become more complicated if we also take into account what the different predator species do to each other. Recently there has been a great deal of interest in the realization that carnivores also prey on other carnivores, and in the implications of this in entire ecosystems: one talks about 'trophic cascades' (Polis & Holt 1992). We have seen already that coyotes may keep numbers of red foxes down, and nesting ducks may reap the benefits from that (Chapter 5). The effects of coyotes also extend to other small predators, such as grey foxes and feral cats. By removing coyotes, populations of these smaller carnivores increased dramatically (a process called 'mesopredator release') in a study in California, causing substantial declines in numbers and even extinctions of breeding birds (Crook & Soule 1999).

In Spain, Paco Palomares and his students showed that in areas where the Iberian lynx still thrives, the animals suppress numbers of the Egyptian mongoose, a major predator on rabbits, and consequently rabbit numbers are higher (Palomares *et al.* 1998). Not surpisingly, therefore, predation by carnivores may even extend its effects to the vegetation: for example, on Isle Royale, Michigan, it was demonstrated that at times when wolf predation kept moose numbers low, the productivity of fir trees was dramatically higher (Post *et al.* 1999). Thus, the top-down effects of large predators may be felt through several levels of the 'feeding-pyramid', throughout the entire ecosystem.

ARTIFICIAL IMMIGRATION: INTRODUCING PREDATORS

Natural immigration of a 'new' species or predator or prey into a community is extremely difficult to observe. However, we can learn a great deal from the usually highly undesirable interference by people with wild animals outside their normal range of distribution. In this section I will look at some cases of 'unintended experiments' that demonstrate some of the effects of carnivores on other species, and the way in which prey accommodate them, or not, as the case may be.

Some time in March 1979, I was sitting against an uncomfortably sharp piece of lava which provided a modicum of shade, surrounded by the most inhospitable landscape imaginable. There was no vegetation, the rocks and sand were black and unbelievably hot, there was a distant volcano on my left and the ocean on my right. Moving about was difficult because of the sharp, un-eroded lava, but in any case I did not want to move because of the heat. I was on the south shore of Isabella, the largest island of the Galapagos archipelago, right on the equator. To me it looked like one of the most hostile places for any kind of animal existence, but nevertheless it provided a home to thousands of marine iguanas and some fur seals. There should be little else, apart from just me and my small tent, at least 60 km from the nearest other human.

But I was there precisely because of a species that should not have been there at all. In front of me was a pack of four feral dogs, some 100 m away, trotting carelessly over lava which I could negotiate only with the greatest effort. The dogs were unaware that they were being watched, which was just as well because they could be very aggressive and I had nowhere safe to go. Suddenly, without warning, the lead dog dashed off to the side, followed almost instantaneously by two others. Fewer than 5 m away they had spotted a marine iguana, a large male well over 1 m long, which they grabbed between the three of them and, in almost no time at all, tore to pieces. It all happened right out in the open, in broad daylight, but the iguana never stood a chance. All four dogs ate, and 15 minutes later they were on their way again, leaving me to study the remains of their kill.

I could calculate the exact size of the iguana from the head that the dogs left behind, and from that I could estimate its original body weight: 5.2 kg. The dogs left 2.2 kg, and therefore ate an average of 0.75 kg each, which is probably about their average feed for the day. I got further data from other sources: there were many scats lying around, since nothing seemed to decompose in the heat, it just dried up. The scats told me in precise detail that marine iguana was by far the most important food for the dogs there. The scene was completed by vast numbers of pitiful iguana remains.

Domestic dogs came to Galapagos in the prison colonies set up there in the late nineteenth century. They became feral there and had since been totally wild for many generations. However, people who knew Isabella had only seen them move into this stretch of coast over the last few years. They were large animals, somewhat like hounds, mostly white with some dark blotches, individuals easily recognizable in this eerie dark landscape. Over the weeks that I was there I estimated the numbers of dogs, and the numbers and size classes of marine iguanas, and I worked out what the dogs were eating. The results were a horrifying confirmation of what I expected when I first landed on the island: dog predation accounted for huge mortality amongst the iguanas, something in the order of 27% of the population per year (Kruuk & Snell 1981).

There is no way that such losses can be sustained. Iguana reproduction is extremely slow, and these gigantic lizards take many years to mature. So the dogs spelled a rapid end for these unique reptiles along the coasts of Isabella, as they did for iguanas in many other places in Galapagos. On the second largest island, Santa Cruz, feral dogs wiped out the entire population of another species, the land iguana, between 1975 and 1977. A similar fate befell land iguanas on the southern part of Isabella. Marine iguanas, which feed in the sea, are slightly less susceptible to dog predation because of their often almost inaccessible cliff habitat. Nevertheless they have been exterminated from many coasts, wherever dogs occurred. Giant tortoises were decimated by dogs on Santa Cruz in the 1970s, and dogs have also caused heavy losses amongst Galapagos penguins and petrels. In response to my dog study, the National Parks authorities took firm action, and exterminated the feral dogs along the southern Isabella coast in a fast and efficient campaign.

Raccoon

This tale about an exotic species is just one of many from around the world, from numerous islands on which prey species have been totally obliterated or decimated by imported predators. Island species are especially vulnerable, because they have not been in contact with predators before. Without doubt, domestic cats are the worst offenders amongst the carnivores. To quote the New Zealand ecologist Phil Moors, 'no other alien predator has had such a universally damaging effect on seabirds'. But there are many other cases of predator introductions, involving two different mongooses, the American mink, ferret, stoat, weasel, red fox, Arctic fox, Patagonian fox and raccoon, as well as the domestic dog (Moors & Atkinson 1984). To crown it all, next to these carnivores many other non-carnivore, vertebrate predators escaped or were released and made their impact on islands, species such as various rats, the house mouse, pig, hedgehog and others.

Each of these predators can have a massive impact on the native fauna, and many of them have completed the process of extinction. Losses of island-breeding seabirds to cats have been described as 'catastrophic', with numerous well-documented extinctions of island populations, even of entire species (Veitch 1985). On Ile aux Cochons, one of the Crozet Islands in the southern Indian Ocean, cats were responsible for the extermination of ten species of petrel (Derenne & Mougin 1976). Such horror stories abound, and many cases must have gone undocumented.

If exterminations are only local, then the seabirds may come back again to islands after the cats are removed, thereby completing the experimental demonstration of the effect of predators. This happened on Baker Island (near Fiji in the Pacific Ocean) where cats were exterminated in 1964, and in 1978 colonies of sooty terns and frigate birds were firmly re-established, having previously been slaughtered out of existence by the cats (Forsell 1982).

On Hawaii, Fiji and other Pacific islands the small Indian mongoose was introduced, for the purpose of controlling rats in sugar cane fields. It caused extinctions of several petrel and shearwater species (Bourne 1965). Red, Arctic and Patagonian foxes have also left a trail of island extinctions behind them, each in its own area, after being introduced for the fur industry. In an interesting twist to this litany of terror, sterilized red foxes were used to exterminate (introduced) Arctic foxes on some of the Aleutian Islands (Bailey 1982).

The most spectacular damage to an entire fauna has been done by predators in Australia, a continent historically without Carnivora. In the relatively short time span of less than a 100 years, 17 species of mammals (6.3% of the mammal fauna) became extinct, largely because of the introduced predators. The red fox was introduced into Australia by farming colonists in the late nineteenth century for sport. Their numbers appear to be mostly dependent on rabbits, and 'predation by foxes is implicated in the extinction and rarity of several marsupials, bandicoots, small wallabies and rat kangaroos...' (Newsome & Corbett 1977). Fox removal, in well-controlled experiments in several areas of Western Australia and New South Wales (by poisoning with an agent called 1080), resulted in immediate and spectacular population increases of several species of rock wallaby, bettong, numbat, possum, tamar wallaby, mallee fowl and others (Saunders *et al.* 1995).

Feral cats have an even worse record than foxes in Australia, especially in desert country and on islands. They are blamed for the continental extinction of seven species of mammals, and regional and island extinctions of many more. Cats had an especially drastic effect on the populations of smaller mammals, those less than 3 kg in weight (Dickman 1996). Between them, cats, foxes and dingoes finished off ten species of medium-sized marsupials, although some of these do still exist on a few small islands off the Australian coast where the predators are absent (Corbett 1995). Six recent attempts to reintroduce marsupial species into areas from which they had disappeared failed because not enough effort was spent in exterminating cats and foxes (Short *et al.* 1992).

In Europe the American mink escaped from fur farms and it was released in numerous places, starting in the 1920s. It is now abundant virtually everywhere. There is good evidence that in Britain it alone is responsible for the sharp decline and likely demise of the water vole, through direct predation (Strachan & Jefferies 1993). American mink are also held responsible for the alarming decrease in numbers of several bird species in Britain, and they have totally exterminated many seabird colonies on islands in Scotland (Craik 1998).

Of course, despite these carnivore introductions many potential prey species continue to thrive in the affected areas. But that does not invalidate the point that several kinds of carnivore can totally obliterate prey species in a very short span of time, caused by thoughtless human interference in the first place. It is only through happy accidents of geography that much of this sort of carnage is prevented.

The results of introductions all over the world, with all their un-intended effects, demonstrate with very little room for doubt that carnivores can cause extinctions (despite the fact that these extinctions were engineered by ourselves). It is more than likely, therefore, that carnivores must also have done so in the past. Not only do they cause extinctions, but by just being in a place, carnivores are likely to have pre-vented populations of prey species getting established at all. A logical conclusion is that on every continent the fauna would have looked very different from what it is now, if there had been no carnivore predators.

In the ecosystems we find today, it is likely that those vertebrate prey species that have survived, are those with at least partly effective anti-predator mechanisms against carnivores. These anti-predator mech-anisms are the fascinating, intricate behaviour patterns that I will dis-cuss in the next chapter. They are present in mankind and in all other species that did not succumb in the confrontation with carnivores.

Black-headed gulls attacking American mink

10

Crying wolf: anti-predator behaviour
Anti-carnivore behaviour of animals and man

Getting close to a wild carnivore can be an emotional experience, whether it be a fox or a lion. One feels that unique adrenalin-based chill in the spine, a slight fear, the overwhelming thrill of the animal's proximity and the knowledge of its abilities, its sharp teeth, speed and its strength, its possible aggression and the question of what it will do next.

Such an emotional response stems from our basic instincts about animals, and particularly about carnivores. These instincts are innate in even the most urbanized of people, and must have evolved to meet the fundamental ecological challenge of predators to our survival. In previous chapters I have discussed details of these threats from carnivores to ourselves and other animals. Here I want to describe the behaviour of a variety of species, including people, that are targets for predators, behaviour that we see in response to such threats.

The ecological effects of carnivores on ourselves, and on almost all mammals and birds have been, and still are, considerable. The next question, then, is about the behavioural response: exactly how do birds and mammals protect themselves? Do they learn how to cope with danger, and/or is there innate behaviour that evolved in response to selection pressures from predators? Seeing the vast range of beautifully adaptive behaviour patterns in animals facing all kinds of environmental challenges, one could also expect many behavioural adaptations to protect against the threat of predation. So, in other words, what armoury of responses do animals have available against carnivores, and what about ourselves, *Homo sapiens*? The further analysis of this has profound implications for our understanding of past and present-day public reactions to carnivores. In order to know our own behaviour, it greatly helps to see first of all exactly how the anti-predator responses of other animals are organized. There is a considerable amount of research about this in birds and mammals, and here I will first describe some bird reactions in detail.

ANTI-PREDATOR BEHAVIOUR OF BIRDS

In the late 1950s the Dutch ethologist Niko Tinbergen was one of the first scientists to be struck by the delicate adaptations of birds in their protection against predation. Later he was to receive the Nobel prize for his outstanding studies on animal behaviour. Tinbergen's work started very simply with field experiments on gulls, where he focused on just one small behaviour pattern: the response of adult birds to shells of eggs in their nest, after their chicks had hatched (Tinbergen *et al.* 1965). In the early 1960s I was fortunate enough to work with him on the project as a student assistant, and it was an experience that had me hooked for life. Tinbergen was a wonderfully stimulating person, a naturalist with the ability to teach. He showed how even the smallest details of bird behaviour could contribute to survival in the struggle against predation.

After the gull chicks hatched, their parents carried the now useless eggshells a long way from the nest. This demonstrably improved the survival of the brood, because having eggshells lying around makes a nest more conspicuous, guiding predators to the chicks. Eggshell carrying is a mechanism of camouflage, to protect against predation by crows and other gulls. Tinbergen showed that the behavioural response, the removal of an empty eggshell from a nest is 'innate', i.e. it does not have to be learned from others or by experience. He also showed how birds distinguish between eggs and eggshells, and how exactly they time their response.

It is only one tiny behavioural adaptation, a very small part of a highly intricate battery of defences against predation, which is called the *indirect* anti-predator system because it is put in place without the predator being present. This security system involves, amongst other behaviours, elaborate camouflage of eggs and chicks. Camouflage only works if nests are spaced out, and the exact site of the nest and habitat selection is crucial. Colony breeding is also important, as birds help protect each other. Many such defences, and colony breeding, work only if egg laying is synchronized.

The birds have many more indirect adaptations to predation. But after my involvement with the eggshell study I became especially fascinated by the birds' *direct* reactions to their enemies, the behaviour of gulls when they were immediately threatened. So it was that I started my PhD study with Tinbergen by analysing the entire range of the gulls' direct responses to the presence of their predators, to see how these defences differed and how effective they were (Kruuk 1964). It was one of the first studies bridging what was then a gap between ecology and ethology.

I watched black-headed gulls that were breeding in a colony of many thousands in the sand dunes of Ravenglass, along the English coast of the Irish Sea. The gulls had a host of enemies, all taking their toll. There were foxes, stoats, badgers, hedgehogs, people, crows, herring gulls, peregrines, hen harriers, short-eared owls and several others. Each of these predators had its own special interest and its own way of achieving its aims: some ate eggs or chicks, some took only adult gulls, and others preyed on both adults and nest contents. Some predators worked at night, others in daylight, some approached on the ground, others from the air. Moreover, some of these enemies made only the occasional attempt on the gulls' lives, whereas others showed an obvious specialization in gull predation. Some of this I could quantify to produce a detailed picture of the threats facing the colonies.

It was just one single case study, but here was a scenario that provided an example for goings-on everywhere in the animal kingdom. There seemed little doubt that between them, the predators would easily have wiped out the whole colony of gulls within a short span of time, if the gulls had not evolved an efficient security system. To protect themselves, the gulls had a large range of reactions to their enemies. Each of these behaviour patterns included a mixture of elements such as *aggression, fear, socializing* (i.e. attraction to other gulls) and something that I simply called *curiosity*. I could recognize these ingredients from postures and movements of the birds, and from the distances they kept from an intruder and each other, and from other associated behaviour. The mix of behavioural ingredients in the reactions of the gulls is different for each type of predator. For instance, a peregrine preys on adult gulls only, and when it turns up it causes almost pure fleeing behaviour in the gulls, pure fear and flight in total panic, whilst at the same time the gulls are highly attracted to each other, flying very close together.

In stark contrast, a hedgehog or a crow in the colony is not interested in adult gulls, but it aims to take eggs and chicks. The gulls respond with pure aggression, unadulterated attack, by dive-bombing and pecking. In reactions to other predators, to species like foxes or stoats that are dangerous for adult gulls as well as for eggs and chicks, the gulls show a mixture of fear and aggression with some other behavioural elements thrown in, such as socializing and curiosity.

These mixtures of behaviour show up clearly when the gulls form a flock near the predator. When the enemy is a stoat the flock stays close to it, whereas when it is a fox the gulls stay far away; this is adaptive, as a fox is more dangerous to adult birds. When the gulls' eggs are hatching the gulls attack a lot, but when the chicks get older the adults attack far less; this is adaptive because larger chicks look after themselves, by hiding. From the behaviour of a flock of gulls I could often recognize which predator was present in the gull colony, even at a long distance (and I believe that if I could do this, the gulls could as well).

Many aspects of these behavioural differences appear to be adapted to optimize the defence of the birds themselves and of their off-spring. Diving attacks, hitting the predator and pecking it may cause it to retreat from the brood, but it does expose the adult bird to a pair of fast jaws. Fleeing protects only the adults themselves. Flocking, synchronizing responses and staying close to conspecifics may confuse the predator on adult birds. An optimal response is a mixture of such behaviours that takes into account the kind of threat that a predator poses to adults and brood.

However, it is clear that the responses of the birds do not fully protect them – in fact they are nowhere near fully effective. They are more effective against some predators (e.g. crows) than against others (e.g. foxes). The predators, of course, have also evolved behaviour patterns to counteract the gulls' behaviour, and many adults, chicks and eggs are taken, especially by other large species of gull and by foxes. These predators often kill substantially more than their immediate requirements, so they have quite an impact on the colony, despite the gulls' defence measures. Nevertheless, there was little doubt that without the gulls' responses the impact would be much more drastic. Without the gulls' own behavioural protection predators would have wiped them out in a very short time.

One important behavioural ingredient of security is *curiosity,* a clear attraction from a potential prey species towards the most dangerous and effective predators, which I will come back to again later as it appears to be very important also in our own species. Birds often came from long distances when there was a predator commotion in the colony, hovering over it, sometimes landing, looking intently at it, without showing much aggression or fear. When standing around on the ground such gulls were often 'long necking', and it was difficult not to compare such a scene with people gathering around an accident.

It seems likely that this curiosity helps the birds to learn what kind of adversary they are facing. There is a need to learn, because the birds need to face strange and new predators, such as an unusual breed of dog, or a ferret or an otter (this last one was very rare at the time of my study). The newly introduced American mink also draws a highly appropriate reaction, a mixture of fear, aggression and curiosity on a par with the behaviour towards a stoat. The birds seem to have an innate predator recognition mechanism that covers all carnivores and many other predators, and they can hone these responses by a learning process, especially by learning from each other through this curiosity.

I first suspected that birds might learn about danger from a particular predator after watching one of my colleagues in the colony, a student who often collected dead gulls and carried them around with him. He habitually caused tremendous scares amongst the birds, huge flocks going up from their nests and flying towards and over him, even when sometimes he went about empty-handed. The gulls recognized him, and their response to him was much more spectacular than to any other person. There was no doubt that he was seen by the gulls as particularly dangerous to the adult birds, and they learned from what the predator had (apparently) done to their conspecifics.

I decided to do some experiments to explore how this could work. I displayed a mounted model of a stoat in a colony of a different species, the herring gull. The model was sometimes on its own, sometimes together with a dead gull (Kruuk 1976a). This was in a colony on Walney Island in the Irish Sea, where stoats did not normally occur.

The stuffed stoat caused great curiosity amongst the gulls, which stood or hovered very close with wide-open eyes. Some birds showed much fear and launched vigorous attacks, just as to a live stoat. The reactions were especially spectacular when I put a dead gull next to the mounted stoat: then the birds responded by flying up a long distance away, and hovering over the model without aggression, showing clear curiosity. Gulls did not respond at all to a dead bird in the colony without the predator model. I could demonstrate that the gulls learned from what they saw, because after the mounted stoat had been in the colony with a dead gull the birds were much more wary of the model even when I put it out on its own, and for a long time afterwards. The observations confirmed that during the strange attraction of the gulls to predators, curiosity about the death of another bird had taught them something about the dangers of the predator.

So in these observations from the Walney gull colony, I found a way in which experience can be passed on in bird society, as the gulls watch what happens to others. Of course, there are other additional behaviour mechanisms too. For instance, most species, including gulls, use alarm calls. These calls are given especially by parents in the presence of chicks, but also by unattached birds, when there is a predator about. One immediate effect of an alarm call is that all who hear it can react to the predator with curiosity, fear or aggression. The alarm call synchronizes behaviour, and that means that the alerting animal is not alone any more; it finds strength in numbers. On top of that, alarm calls may also pass on predator experience from parents to offspring.

Interestingly, the alarm calls of some individual gulls were usually ignored by their neighbours. These individuals were birds that 'cried wolf' too often, nervous animals that saw danger when there was none. The other gulls just could not be bothered to react any more to them, but if one of the more 'reliable' gulls cautioned with an alarm call, the whole neighbourhood would fly up around them.

What I have described here is the response of just one group of birds, the various species of gull, but we can see the basic principles of their anti-predator system in many other animals. There is the *indirect* protection, through living in herds or flocks, synchronizing their breeding, or through camouflage, by selecting the right habitat and so on; and there is the *direct* defence. All animals show a different mix of reactions, but the same ingredients are there: the aggression, fear, curiosity, attraction to others of the same species in the presence

Stoat

of danger, alarm calls and, in many species, the ability to learn from the mistakes of others.

ANTI-PREDATOR BEHAVIOUR OF HOOFED MAMMALS

In order to see mammalian reactions to their predators, we can easily observe wildebeest, gazelle, topi, hartebeest and zebra mixing on the open grasslands of East Africa, where they are exposed to a range of different enemies. At any time whilst the ungulates are grazing a cheetah, lion, hyaena, leopard, wild dog or jackal may show up. The reactions of the herbivores to such a predator have been intensively studied, and over the last 30 years many scientific papers have been published on the subject. One, not very surprising general conclusion is that the more a particular herbivore species occurs in the diet of a carnivore, the more fear it shows for that predator. For instance, gazelles flee a cheetah at a much greater distance than do wildebeest, and wildebeest in turn flee earlier than do zebra (Kruuk 1972a).

Moreover, the more dangerous a particular carnivore is for an ungulate, the more curiosity it releases. The potential victims, and this probably includes all species of herbivore out on the open grassland plains, often approach a predator (although keeping a given minimum distance) and stare at it, heads held high, sometimes uttering alarm calls (Walther 1969). This curiosity is immediately followed by fleeing as soon as a predator gets too close. On the Serengeti plains it is an unforgettable scene to see whole herds of several different species all staring at a quietly walking large cat, like a lion or cheetah. They may follow it, and one cannot help but compare such a herd to a crowd of people, gaping at something or somebody.

A Cambridge PhD student in the Serengeti, Clare Fitzgibbon, spent several years on a fascinating study of the behaviour of Thomson's gazelle or 'tommies'. She showed that curiosity from potential victims does affect predators. Cheetah, for instance, are less likely to attempt a capture, and move on further, after being stared at by tommies. But, most importantly, Fitzgibbon demonstrated that the tommies' staring also carried a small risk: it costs grazing time, and it costs lives. For young gazelles one in 417 episodes of staring at a cheetah ended in death, whilst adult gazelles could stare with somewhat greater impunity: one in 5000 were killed (Fitzgibbon 1994). Fitzgibbon argued that at such risk the tommies must be getting something worthwhile for their curiosity.

Most likely, the gazelles had the potential to obtain information about the predators' tactics or strategies. The poet W.H. Davies, writing in 1911, obviously referred to man but I think that his poetry applies equally to animals when he exclaimed 'A poor life this, if, full of care, we have no time to stand and stare' (Quiller-Couch 1949). Contemplation enriches human minds and saves ungulate lives.

There are some general aspects of anti-predator behaviour that differ greatly between birds such as gulls, and mammals such as the African ungulates. Mammals respond to much more subtle details in the behaviour of their enemies than do birds. For instance, I was able to see the full range of the gull and other bird anti-predator responses by presenting them with a stuffed, mounted fox, or a stoat or a crow, or just a cardboard silhouette of an owl or a bird of prey. But Serengeti wildebeest and zebra were fooled less easily, and their reactions were much more discriminating with respect to the behaviour of a predator. These animals responded to a stuffed hyaena not much more than they did to a wooden log.

For my African experiment I had gone through great trouble to acquire a mounted and stuffed hyaena, and I thought it was a beauty. But the wildebeest just seemed to laugh at my experiment, and totally ignored it. Later I realized that these animals react to a host of small signals from the predator, to how it walks, which way it looks, whether it has just eaten, and more. It taught me that since I can see from looking at a lion or hyaena whether it is likely to hunt or not, then the wildebeest or zebra can do so at least as well as I can, and they respond accordingly. This is not surprising perhaps, as after all it is their survival that is at stake. Nevertheless birds do not seem to have reached that level of sophistication in their security system.

It appears likely that this ability of ungulates to recognize the probability of attack by the appearance and behaviour of a predator has been acquired by experience, and that perhaps the animals learn from each other's reactions; they may also lose the ability to recognize danger if a species has not been exposed to a particular predator for several generations. This became clear when brown bears recolonized Scandinavia, where they had been wiped out a century earlier: they were able to kill elk (moose) with unusual ease. The elk were totally oblivious to the threat of bears, but within one generation they were back to 'normal' again, and bears had the greatest difficulty in killing them (Berger et al. 2001).

As with gulls near a fox, so with wildebeest and gazelles that walk in a posse behind a cheetah, or with songbirds around an owl, or birds chased by a peregrine: there is one aspect of anti-predator behaviour that one sees almost everywhere, and that is a strong attraction to other members of the same species when there is danger. One might call it a 'common-enemy phenomenon', a united-we-stand, a finding safety in numbers. Sometimes, the benefits of such tactics are immediately obvious, for instance when hyaenas or wild dogs chase a single wildebeest across the plains. Once the intended quarry manages to rejoin a herd it just seems to disappear: the dogs or hyaenas lose it and usually the hunt is over. The other side of the coin is that the predators clearly try to prevent this, by separating a single quarry from a flock or a herd.

Another very interesting phenomenon shows up in anti-predator behaviour when we also take the social behaviour of the prey animals into account. This was first shown in fish, in sticklebacks, by the animal behaviourist Felicity Huntingford. She found that those species and individuals that are more aggressive to predators, such as a pike, are also more pugnacious to their own conspecifics, to other sticklebacks (Huntingford 1976). It means that the mechanism of anti-predator behaviour of an individual shares important elements with the reactions to its own species. The responses are affected by hormones, not just in fish but in many species. Endocrine mechanisms explain why many characteristic anti-predator behaviours occur especially during the reproductive season. The upshot is that some individuals or categories of individuals, for at least part of the time, react more aggressively to or fearfully of their own species, as well as to predators.

In summary, in many and possibly all mammals and birds we find a mixture of fear and aggression in their reactions to their predators, a mixture that is usually well adapted to the kind of immediate threat posed by the enemy. The fear and aggression are often supplemented by curiosity, especially to the most dangerous foes, and it has been shown that through this curiosity the animals are able to learn about what happened to others. Almost all species have some kind of alarm system. An animal which is faced by an enemy also often shows an attraction to members of its own kind, a 'safety in numbers' response. The reaction to predators may be affected by hormones which also influence the behaviour to conspecifics.

RESPONSES TO PREDATOR COMPETITORS

So far I have focused on the reactions of animals to their actual predators. Unfortunately, the waters are muddied by another complex issue, by competition. It is an issue that becomes particularly important when later in this chapter I talk about carnivores and man. In many cases one species is an enemy loathsome to another, not just because it is a predator, but because it also competes for resources. A good example is the carnivore 'guild' on the Serengeti plains, where lions kill cheetahs or hyaenas or jackals. Lions act as predators on these other carnivores, but they also compete with them over carcasses, and they catch similar prey. Hyaenas also compete with and sometimes prey on these other species (Kruuk 1972a). Competition with lions and hyaenas is one of the main causes of local extinctions of the African wild dog. This process is threatening the survival of the species, which is now confined to relatively small areas of a fragmented habitat (Vucetich & Creel 1999). There are numerous combinations of carnivore species where we find the same predator-cum-competitor relationship, such as with honey badgers and jackals, wolves and foxes, and foxes and cats.

When, between species, we have a competition-only situation, then the usual interspecific behavioural response is aggression, and this accounts for many aggressive interactions between carnivores. The closer the competition, the more aggression the other species evokes. A simple, untrammelled case in birds (which is simpler because there is no predation involved) is that of vultures on a carcass. For instance, in the Serengeti up to six species of vulture may feed together on a predator kill, with just straight competition between them for juicy chunks of meat. The different species prefer different parts of the carcass, and the closer their interests tally, the more they fight whenever they meet (Kruuk 1967). There is no predation between the birds, only competition, which leads to aggression.

However, when carnivores or other hunters react to each other there is a combination of predatory interest, of fear of predation, and of competition. Competition is likely to increase the element of aggression in their reactions to each other, which is a very relevant point also in the relationship between carnivores and man. What is important is the strong element of ambiguity in these relationships. If one species preys on another, the risk can be minimized by fleeing; if the two species compete for resources, they can try and displace each other through aggression. But if the two are both competitors as well as in a predator–prey relationship, then the behavioural response needs to be a compromise, which by its nature is bound to be less than fully effective.

ANTI-PREDATOR BEHAVIOUR OF MANKIND

The observations on birds and mammals that I described in the previous sections, as well as others, suggested that the basic make-up of the security system or anti-predator behaviour is present in many different birds and mammals. From many casual observations I think that the basic components are probably there in all species, with comparable direct anti-predator behaviour in their strategies if not in their tactics. Of course, the details of the anti-predator system are different in every single species. But the fundamental ingredients are probably the same, and I think that they are also deeply engrained in our species.

Obviously, we cannot just extrapolate from wild animals to people. As we saw above when looking at the anti-predator behaviour of wild ungulates and gulls, there appeared to be a large leap in behavioural complexity from birds to mammals, such as that expressed in the different reactions from birds or wildebeest to the stuffed stoat and the stuffed hyaena. If one were to extend comparisons even further to people, a quantum leap in sophistication can be expected.

In the anti-predator behaviour of mankind, social learning and cultural transmission are especially likely to play a more prominent role. Nobody would doubt that we rely much more on what we learn from others than on our 'instincts' in comparison with mammals and birds. Nevertheless, the same basic principles such as the importance of fear and aggression, the roles of curiosity, and the attraction to conspecifics when under threat, all play a role.

The evidence for this is fairly persuasive, despite the fact that carnivores play much less of a role in our lives than they do in the existence of wild animals. When I describe the reactions from wild prey species to different kinds of predator, as I did above, the generalizations about fear and aggression may sound unsurprising to us. They sound predictable precisely because our own human responses are the same as those of the animals: they are 'common sense'. If a predator threatens your child you go for it (aggression), but if it is only likely to attack you yourself, then you try to avoid it, you flee. Similarly, human curiosity when confronted with dangerous carnivores, and our attraction to other people when there is danger, appear to follow some of the same rules as found in wild animals. Thus, dangerous predators cause similar kinds of overt behaviour in us as they do in wildebeest, zebra or gazelle. This is not to state anything about the underlying emotions of the people involved, but merely describing the objective, observable behaviour of mankind in the presence of danger from predation.

A specific example is the behaviour of people to one of their classic predators, the lion. Mostly, our instinctive reaction during a confrontation is one of terror and fear. When visitors on bush-walking safaris are taught what to do when walking in wild country in Africa, they have to be persuaded hard not to run away when a lion approaches, but to stand their ground or climb a tree. It is very difficult to get people to do this because of the almost irresistible urge to flee, which could be fatal, because for many predators the sight of a fleeing individual is an added stimulus to hunt.

We also know that some people show aggression to lions and want to kill them: witness the ritual spearing of lions by Masai, and the wealthy white hunter spending a fortune on the pursuit of the king of beasts. Just as in other animals, in humans these basic urges of fear and aggression towards lions are also accompanied by what is another form of attraction, i.e. curiosity, by the appeal of danger. A television programme on lions is always popular, and lions are the favourites of visitors to African national parks. In the safety of a vehicle, people drive as close to these animals as they can.

Another symptom of the appeal of danger from carnivores appears in our newspapers, whenever a carnivore does something untoward to man. Typically, in January 1998, the story of a circus tiger seriously wounding Mr Chippendale in Florida got front-page coverage in many papers throughout the world, but the unfortunate trainer would not have had a mention if he had been hit by a car. Similarly, a jogger killed by a cougar or a hiker killed by a bear in Canada makes newspaper headlines.

This appeal and attraction of carnivore danger is obvious not only in a zoo, where children and adults are drawn to the lions, tigers and wolves as to a magnet. The extra attraction of carnivores, especially cats,

Lion in zoo

is very striking (Balmford *et al*. 1996), but it is also part of a more general phenomenon, and in just the same way people in a snake park flock towards the most poisonous inmates, the cobras and mambas. Apparently, we have a great need to stare at the tools of the grim reaper.

Analysing the attractiveness of the dangerous carnivore somewhat further, we find that there are at least two separate aspects that draw people. Firstly, we are drawn by violence, by threats to survival in our own society and in that of others: witness the general appeal of violence in movies and on TV. There seems little doubt that this human behaviour is comparable to the curiosity aspect of anti-predator behaviour in animals. We are interested in the mechanisms of danger, and the fate of the attacked.

Secondly, hunting fascinates for some other reason. Whether one is excited by it or repulsed, whether one feels compassion for the prey or admiration for the predator, it is an aspect of animal behaviour that affects us deeply. We identify with the hunter, perhaps because of our own hunting instincts; if we had been mere herbivorous plodders at heart, I do not think that the sight of a cheetah would have affected us in quite the same way.

Can we generalize from observations such as those of man's direct response to a large predator? Can we extrapolate from human behaviour when faced with a lion, to our attitudes towards carnivores in general? Or, putting it differently, if we know something of what people feel about lions, does it tell us anything about their reactions to badgers, bears, foxes or weasels? I think the answer to this is a qualified 'yes', and to argue this, I again fall back on studies of animal behaviour.

If a ground-nesting bird such as a plover, a goose or a gull sees one of its eggs outside the nest, it will roll it back in. As the ethologist describes it, the egg provides a stimulus for the bird to perform egg-rolling behaviour. Give this bird two eggs outside its nest, and it will show which of the two it prefers, by rolling that one in first. If, for instance, one of these eggs is its own, natural size, and the other one is twice as big, the large one will be retrieved first, and much more enthusiastically (Baerends 1970): it may be quite unnatural, but it provides a *supernormal* stimulus. Similarly, very small eggs may be recognized as eggs despite their size, but the response is half-baked. Wildebeest show curiosity and some aggression to a serval cat, despite its minute size and irrelevance as a predator, since it looks like a very small leopard. Sheep also respond similarly to a domestic cat, as do many birds and mammals to a weasel. The objects of these interests are not predators but they look like predators, in a pocket-sized version.

The writer and student of human behaviour Desmond Morris argues that for people the same principles apply, and we find supernormal (and subnormal) stimuli everywhere (Morris 1967). If you respond to something in another person, then you are likely to respond even more so when the stimulus becomes exaggerated. That is why fashion makes lips redder, eyelashes longer and busts larger, making them just that much more attractive, and why longer canines scare one – hence Dracula. It seems that we have a strong instinctive response to the killer attributes of large predators. To smaller carnivores, with smaller canines and claws, we respond with the same ingredients of a behavioural reaction but in a scaled-down version. This response can then be modified by learning, but there is always the underlying presence of an instinctive reaction. We treat the small carnivores a bit like the large ones, and this way the sins of the large villains are visited upon the innocent lookalikes.

Moreover, just like other species we also show a behavioural response to carnivores as competitors, as animals that take the quarry we would like to hunt ourselves, and that take our livestock. Predictably, we respond with straight aggression, with a strong urge to kill the perpetrators. This aggression too, may be reflected in our behaviour towards lookalikes, to other carnivores that are harmless but just happen to look like those that compete with us. In Africa it happens, for example, to aardwolves or bat-eared foxes, which are harmless insect eaters, but are often persecuted because of their similarity to jackals

The main point is that, probably largely instinctively (i.e. without having learned this), we have a set of reactions to a group of stimuli that spell 'carnivore' (and other sets of reactions to, for instance,

Aardwolf and termites

stimuli that characterize 'snake' (Morris & Morris 1965)). The strength of these reactions partly depends on the strength of the stimuli, such as the size of the beast. This behavioural mechanism is far from surprising. Like other species, we have suffered predation by carnivores over many generations, and we have evolved responses. Most probably, these are generalized responses to predators, to the animals that prey on ourselves, and to the competition that kills our livestock and game, and the generalized response is modified for individual predator species.

One of the highly important differences between anti-predator reactions in people and those of other animals is the occurrence of altruism. For instance, people will frequently go out of their way, even to the extent of endangering themselves, to kill a predator which is not a threat to themselves or their kin or their livestock, but a threat to someone else. There are the cases of district commissioners shooting maneating tigers or lions, the many vermin control officers employed in different countries, and each of us would go to the aid of a defenceless person when we see her or him attacked by a predator.

As we have seen, the general behaviour of vertebrates in reactions to predators may be a largely instinctive fear, with aggression, attraction to conspecifics and curiosity, and with these same elements also in the behaviour of ourselves. Perhaps the most interesting part of anti-predator behaviour is that strange appeal, which I have referred to as curiosity; this is the non-aggressive attraction that carnivores have for birds and mammals, which enables a potential prey to learn about danger and is especially important in ourselves. Like other animals, people are fascinated by danger and this shows, amongst other ways, as 'culture', a social learning pattern that is much more important than it is in the behaviour of the gulls or the wildebeest. As I discuss in the next chapter, everywhere, even in our literature, in children's stories, in art, in our superstitions, in coats of arms and other small symbolisms of day-to-day life, carnivores play a role that is more prominent than the role of most other animals. In biological terms, it means that there may be a survival value in this aspect of our culture. We are teaching others what is lethal in the environment, how its deadly forces work, and one might call it a cultural alarm system. The attraction of our species to carnivores expresses itself in this culture, in our admiration for hunting, and in our appreciation of beauty of these wild animals.

Little Red Riding Hood

11

Carnivores in culture
Carnivores in fable, religion, art and heraldry

One might expect the involvement of a people with any one particular aspect of their environment to be evident from their writings, stories and art. Fishing nations, farming countries or hunting tribes all testify to this. If we are looking for a measure of the involvement of mankind with a particular group of animals, we should investigate their occurrence in our cultural expression. This is all the more interesting in the case of carnivores, when such expression may be used to teach our kin about the hazards of life.

CARNIVORES IN CLASSICAL LITERATURE

Centuries ago, during the Middle Ages and in one of the parables that was customary at the time, the topic focused on the court. King Noble the Lion, it was written, decrees that there shall be no more aggravation between his subjects: there shall be peace for everyone. But alas, alas, Reynard the Fox continues his evil machinations, attacking all and sundry. Hersent, the beloved wife of the courtier Isengrin the Wolf is raped by Reynard, after he persuades her husband to become a monk because the food is so good in a monastery. Surely justice will prevail when Reynard is summoned before King Noble to answer for his misdeeds? Alas again, justice there is none, and the murderous villain emerges victorious. The low, cunning Reynard retires to his castle, having left a trail of damage and indignation amongst the loyal citizens (Varty 1967). This tale is from the year 1175, the author is Pierre de Saint Cloud, and his story is an extract of *Ysengrimus*, one of many sagas of that time where animals, especially carnivores, play leading roles. The subjects may vary, but the storylines are not all that different from many a present-day crime series. But in important medieval literature in Latin, French, Flemish, English, German and Italian, animals were often lead characters, and the success of these stories shows in the numerous imitations, translations and variations on the theme. Noble, Isengrin and Renard are joined by Grimbert the badger, Tibert the cat, Brun the bear, Chantecler the cock, Bernard the donkey and others.

The upsurge in carnivore actors on the literary scene started in the middle of the twelfth century, but much of the writing at that time was based on earlier poetry, from ancient Greece or Rome. In those earlier poems, however, none of the animals had proper names. The first medieval writing on animal characters was light-hearted, designed to please and to entertain, but gradually over the following centuries it gave way to heavier, moralizing and didactic allegory. The adventures of the animals reminded people of the high jinks of royalty, priests and other dignitaries.

Reynard became the villainous hero, evil in person, the all-time hypocrite and deceiver, a symbol of sin, the very devil in disguise. Noble the lion on the other hand was royalty, above it all, somewhat dim but authoritative. Isengrin the wolf was a worldly and corrupt monk or priest, a greedy, dull-witted criminal at heart but nevertheless an important figure in society. Brun the bear was the lumbering dimwit who managed to combine power with being the butt of everyone's mischievous practical jokes. Chantecler the cock was the standard, upright and rather simple citizen. Over the years, the Reynard-type story became a wonderful and safe outlet for political satire, translated or retold in all major European languages. In England it was *History of Reynard the Fox*, in France *Roman the Renard*, in Germany *Reinhart Fuchs*, in Holland *Van den Vos Reinaerde*, all of them major contributions to literature in the twelfth to fifteenth centuries.

Many other literary outpourings about animals are found in illuminated manuscripts in Latin, often called 'bestiaries', from all over Europe (e.g. White 1976). They obtained much of their inspiration from Aesop and Pliny as well as from the common, everyday fables and fairy tales of the day. An unknown Greek author from Alexandria, referred to as 'Physiologus' (meaning naturalist), produced a text consisting of the 'facts' of natural science in the second century AD. His text consisted of 48 sections, each about some animal, plant or stone, and was the main source for a vast number of early medieval manuscripts (*Encyclopaedia Brittanica* 1998). There is a wealth of observation in these texts, of animals as well as of people, of animal characteristics in humanity and vice versa, of animal traits that are relevant to us, and of animal habits and their dangers; and very conspicuously, many of the animals involved are carnivores.

The early writings and oral traditions had far-reaching effects on people's attitudes to the world around them. Few people, and certainly no child or adult living in a town, would ever see a wild wolf in their lifetime. But after these stories every soul would know of the wolf's evil greed, the fox's misdeeds and the lion's might. Children were warned to steer clear of dark forests as they were full of ravenous wolves, and the only good carnivore was a dead one. In literature the fox, the cat, the badger, the wolf and the lion added spice to our culture, a spice that very much affected our perception of the real-life predators. In biological terms, details of the lives of the fabled characters were used as illustrations of our anti-predator strategies, when they were transmitted to our children.

Even earlier than Physiologus, Aesop is celebrated worldwide as the originator of the many fables of animals amongst the classical Greeks, but he was an imaginary author – there is no evidence that he existed as a person. The first traceable compilation of Aesop's fables was produced by the Greek writer Phalareus in the fourth century BC, about a century after Aesop was supposed to have lived. Although the collection was lost again some 1000 years later it had a large influence on subsequent literature. About 200 Aesop's fables are known, and even by just glancing through them one notices the predominance of carnivores (Lenaghan 1967). Carnivores are mentioned 83 times in the titles alone, with the most common one being the wolf, which is significant as wolves were the main predators on people in the Europe of those times. In frequency of quotes the wolf is followed by the fox, then the lion and the dog; the cat, tiger, panther and bear also have a look in. Interestingly, some of these animals did not even occur whence the stories hailed, but obviously their exploits were relevant to the teller of the fable and to eager listeners.

From fables an entire style of European literature developed, which subsided again in the late Middle Ages. But the actual fables themselves never gave ground, and they are still around us in a rich proliferation. We find them, for instance, in the seventeenth century in very much the same form as the original Aesop's fables, this time written by the French author and poet La Fontaine (La Fontaine 1997), still frequently rewritten and reprinted now.

Alongside all that there is the somewhat different and more popular form, the fairy tale, the mother's-knee story for children, the tale that is slightly less moralistic than the fable, but not much different and in its own way still full of wise lessons. In Europe, the best-known and most complete compilation of fairy tales originates in Germany. It is the *Märchen*, the result of a lifelong collection effort in the German countryside in the early nineteenth century, by the two brothers Grimm (Grimm & Grimm 1957). Many of their stories have become all-time classics. Who does not know of Little Red Riding Hood? The tale is about a huge, black forest, and the wolf that ate the grandmother first, then enticed the little girl close and swallowed her (whole, just as he had done with her grandmother). The evil beast then fell into a sleep so deep that a hunter could slice open the wolf and replace the uncomfortable women by stones. The wolf drowned because he was so heavy, and Little Red Riding Hood lived happily ever after.

The Grimms' *Märchen* are mostly about people, rather than animals. Nevertheless, wild animals feature frequently, and in 86% of the stories where they occur these animals are carnivores such as wolves, foxes, cats, lions and others. The stories are effective, too. I would guess that the Little Red Riding Hood tale alone has had more influence on popular opinion about wolves than years of scientific research. In children's minds the fabled wolf is usually pure evil: wily, terrifying, hellbent on eating people and set against a background of dark forests.

It is perhaps significant that, at least in Europe even these days, the fairy-tale image of wolves is somewhat closer to the truth than the picture painted by some conservationists over the last 30 years. Scientific information has come mostly from observations in America, where no reliable and detailed evidence ever emerged of wolves preying on people. Some (I think misguided) conservationists therefore deny the danger posed by wolves. But as we saw in Chapter 4, in the wilds of Europe and Asia the wolf hazard to people was very real and still exists today, with the animals posing a considerable and life-threatening danger to the smaller members of society.

In other continents, there are other animals and other stories. Africa has been very prolific in producing animal lore, and there are several excellent collections of animal stories, fables or fairy tales. The actors, the scenery and the belief in the surreal may be different, but the storylines are just as extraordinary as in Europe.

A hunter goes off into the bush, and there he collects two lion cubs, to use as a sacrifice at the naming ceremony for his newly born son. He meets a jackal, to whom he gives one of the cubs to eat, and the jackal promises him a good turn. The other cub is sacrificed, but the cubs' mother finds out and swears revenge. She changes herself from a lioness into a beautiful girl, and persuades the hunter to go with her to her people in the forest. Once in the woods she changes back into a lioness, and summons all the animals with a mighty roar. The hyaena suggests they all eat the hunter, but the jackal pulls a trick, enabling the hunter to shoot the lioness with his bow and arrow. All the animals disperse, and the hunter lives a happy man. This fable originates from the people of the Hausa tribe in Nigeria (Johnston 1966). They are mostly agricultural people, surrounded by the dangers of the African bush. In a collection of 45 of these Hausa stories dating from the late nineteenth century, mammals feature as central characters in 28, and 20 of these characters (or 71%) are carnivores, with lions, hyaenas and jackals being by far the most common. Undoubtedly, in the telling of such a tale there is ample opportunity to weave many wise lessons into it, about what to do and not to do with dangerous animals, and about which ones you can trust and which ones will devour you.

However, it is fascinating that in those and other West African fables the most common central character, apart from man, is not a carnivore, but the spider. It is clever, witty and always comes out on top – the animal one would like to identify with. Jackals are crafty little criminals, hyaenas are greedy, very stupid and always get caught or killed, lions are the dangerous big players who may or may not get their comeuppance. It is the spider that plays the role of the common man. Elsewhere in Africa that role is taken by the hare: carnivores are just as important there, but the scenes revolve around the hare as an innocent but very clever and sympathetic creature. You hear the story, and you know immediately that you can identify with the hare or the spider.

As an example, in the fables of a Tanzanian hunting tribe, the Watindiga (Kohl-Larsen 1956), many more different species are featured than in the Hausa tales, but the hare is the hero. The Watindiga are frequently out in the bush to hunt for a living, and they certainly know their animals, probably much better than the people from agricultural tribes. Out of 67 Watindiga fables mammals are central characters in 49, and of those 22 (45%) are carnivores. They are mostly lion, hyaena and leopard, the most dangerous ones encountered in the bush. The hyaena and the lion play the same roles here as in stories from elsewhere on the continent: the hyaena is contemptible, the lion dangerous and powerful.

It is interesting to compare figures on frequencies of occurrence of animals in stories, with what we know of the present fauna of Africa. In a guidebook to the larger mammal species of Africa (excluding rodents, insectivores and bats) only 26% of 278 mammals are carnivores (Kingdon 1997). More importantly, in terms of actual numbers of individual animals occurring in the bush, the presence of carnivores would be much less significant still, because there are always far fewer predators than herbivores. It means that just as in Europe, the African fables, whether from West or East Africa, are overwhelmingly selective towards carnivores, and especially, of course, towards the large carnivores.

WITCHCRAFT

The African folk tales are only a small step removed from a scene that has much deeper and uncomfortable implications: witchcraft. It is an aspect of human culture that still has a profound impact, especially in many countries in Africa. Once again, carnivores play a vastly more prominent role than any other group of animals. I first became aware

Spotted hyaena

of this in Tanzania, where, despite official denials, witchcraft is still widespread amongst rural populations. Even many people with an extensive formal education sometimes profess to be deeply affected. They may be stuck for rational explanations, but they have no doubt about the penetration of witches into society, and their profound influence over people's lives.

The animal involved in witchcraft more than any other is the spotted hyaena, a species which generally is utterly loathed throughout the continent. I will quote from my own earlier study (Kruuk 1975):

> This loathing goes beyond feelings based on mere ecological competition; it may well be that a primitive fear is involved, arising from the knowledge that hyaenas are 'the living mausoleum of the dead', as someone described them. Aren't the animals' weird laughing noises and its slinking nocturnal movements around one's house (often followed by some disaster to the occupants) almost proof that in some devilish way it is under control of supernatural powers?
>
> Undoubtedly, hyaenas play a more important role in African witchcraft than any other animal. Everybody is aware that people known to be witches ride hyaenas at night (that is why hyaenas' backs are sloping), laughing madly, while casting their spells on other people. Not only do witches ride hyaenas, they also keep them at home, and they live off hyaena-milk and hyaena-butter, and use this butter to fuel their torches. According to some villagers one can smell for days the places where witches spilled burned hyaena butter from their torches (this, incidentally, is exactly the smell of hyaenas' anal glands, used for scent marking). One can protect oneself and one's cattle against witchcraft by feeding the animals ground-up hyaena skin, genitals or heart, or smearing those substances into small cuts in one's own arm. No-one should kill a hyaena, otherwise the witch-owner will take revenge. All this may sound weirdly out of date to a non-African reader, but as late as 1971 I saw in the centre of the town of Musoma a hyaena killed by a car, and in a matter of hours the corpse disappeared, cut up and divided into very small pieces by the townspeople, to be used as charms.

The role of hyaenas in African witchcraft is well documented for several tribes. Some other species, in order of importance, are lion, leopard, snake, frog, jackal, dog and cat. The loathing of hyaenas spilled over even to Europe (where the species only occurred several millennia before), and where in medieval times it was asserted that hyaenas were monsters that never made it to Noah's Ark (because they were 'dirty brutes'). They were said to have come into being after the Deluge, as a cross between cat and dog, killing people and animals at night after immobilizing them by circling three times around the quarry (White 1976).

We also have a good record of the misery caused by African witchcraft when it used the lion as its terror species. This activity must have caused hundreds of deaths. Best documented are the events in the late 1940s in Tanzania with the case of the 'lion men of Singida' (Wyatt 1950). In 1946 reports started to come in to the British colonial administration of several people killed. This was first thought to be due to maneating lions, but gradually it emerged that the deaths were caused by 'wabojo'. Wabojo were 'lion-men' for hire, who dressed in lion skins, killed with knives, and often ate parts of their victims. They were the African equivalent of the werewolf, or one could call them 'werelions'. By early 1947, some 103 people had been killed near Singida, but the authorities found it impossible to get their hands on the culprits themselves, as people were too terrified to give them away. Several people were convicted of hiring the wabojo and condemned to death. Mostly they were women who paid lion-men to kill female rivals or their children.

The case of the lion-men of Singida does not stand alone, but is typical of many incidents throughout the continent, recorded from Angola, Congo, Malawi, Botswana, Tanzania and other places. To protect themselves, people in several parts of Africa were known to eat parts of lion (as I did myself, and I have never been attacked or bewitched since then). People may wear lion charms, such as claws and teeth, which protect and give one something of the strength and courage of these animals.

Witchcraft in Europe, fortunately, is long past its heyday. We will never know exactly what actually happened, but stories abound, and they provide valuable insights into what went on in people's minds. Cats, and sometimes dogs, played a large role as witches' familiars, resulting in the ghastly medieval persecution of cats that I have already mentioned. Inevitably there was also the werewolf, man turned beast, preying on its human victims. There were many similarities with what we still see in Africa today.

RELIGION

In striking contrast with the relevance of predatory animals in witchcraft, and unlike what one might perhaps expect, religion plays almost no role in the relation between people and the dangerous animal aspects of their environment. For instance, Christianity takes remarkably little interest in animals. There are some references in the Old Testament of the Bible, but after the mention of God's creation of animals, and their naming by Adam, such references are few and far between. In the the New Testament of the Bible, there are only a few mentions, with the exception of lambs.

The Old Testament refers occasionally to lions, relating to the days when these animals still roamed the lands of Israel. Significantly, lions usually do not feature as a menace, either as predators or as instruments of the devil, as one would expect. On the contrary, 'the righteous are bold as a lion' (Proverbs 28:1), and 'A lion…is strongest among beasts, and turneth not away for any' (Proverbs 30:30). In other books of the Old Testament we find 'what is stronger than a lion?' (Judges 14:18) and 'Like as the lion and the young lion roaring on his prey, when a multitude of shepherds is called forth against him' (Isaiah 31:4). But the New Testament states 'your adversary the devil, as a roaring lion, walketh about, seeking whom he may devour' (1 Peter 5:8). Wolves, however, are shown as sheep predators, which also attack man: 'ravening the prey, to shed blood', (Ezekiel 22:27), whereas dogs are unclean, fierce and cruel, with which you would insultingly compare your enemy (Psalms 22:16; 1 Samuel 24:14; 2 Samuel 16:9).

Later, in Christian churches, predatory violence occasionally penetrated into the decorative arts, with much symbolism attached. For instance, in Italy in the thirteenth century, the father and son team of Nicola and Giovannai Pisano created for churches sculptures of lions devouring deer or horses. It is argued there that these sculptures have a meaning well beyond first appearances: they are not just an assertion of cruel strength, but they may show the power of the soul, triumphing over the feeble apathy of the body. In the words of the art historian Kenneth Clark, 'the devouring lion, however Christianised, cannot altogether be dissociated from the sacred lions of Persia and Mesopotamia, who owed their sanctity to their strength' (Clark 1977). Clark also quotes William Blake, who in the eighteenth century wrote 'the wrath of the lion is the wisdom of God' and perhaps that statement says as much about our basic feelings towards lions as about the perception of the deity.

References to animals are also uncommon in the other major religions of the world, but they do occur, sometimes in unexpected places. Shintuism, for instance, has the Chichibu Wolf shrine, a well-attended temple in the Chichibu Mountains, at about 2000 m altitude in central

Japan. There are large wolf statues at the entrance, with many decorations of wolves inside. It is the place where about a 1000 years ago a famous local warlord lost his way in the mist in the mountains, and two wolves showed him the way back. He erected the shrine in gratitude. The wolf is now extinct in Japan, but when it still occurred there the locals liked the animal because they felt that wolf predation on sika deer and wild boar prevented crop damage, and as the people did not hunt they did not see the wolves as competitors. Attitudes have changed now, and many people in the Japanese countryside oppose the idea of the reintroduction of wolves, because of fears of predation on livestock (Koganezawa *et al.* 1996).

Similarly peripheral were carnivores in other early religions. To ancient Romans the twins Romulus and Remus were known as sons of the god Mars. Having been abandoned in the wilds by a scheming uncle, they were suckled by a she-wolf and subsequently fostered by Faustulus. Later in life the two brothers quarrelled over where to site (what was to become) Rome, so Romulus slew Remus and then had

The war goddess Sekhmet, Egypt, c. 1300 BC

the town named after himself. One of the themes of that story also occurs in the well-known work by the nineteenth century writer Rudyard Kipling, who had Mowgli nursed by a wolf (Kipling 1994). Born in India and beginning his working life there, Kipling would have drawn this theme from Indian mythology, where it had appeared several times before.

Carnivores do have a place here and there in eastern religions, in the ancient Egyptian beliefs and in Hinduism. The jackal-headed god Anubis accompanied the Egyptian dead into the underworld, and there was Sekhmet, the lion-headed goddess of war and disease, the goddess-lioness Pakhet, 'the rapacious one', and Bastet, the counterpart of Sekhmet, who was a cat. But these were only few in the midst of a plethora of other gods, and even in their images in Egyptian art their human aspects were more important than the animal side.

The Egyptians did get some, albeit meagre and rather vague, practical advice on their environment in one of the first of their Divine Commandments, by being told 'Be thou aware of the lion!'. Several animals were revered by the Egyptians directly, not as gods but as themselves, as animals. Most prominent were the bull and the ram, probably because of their role in farming, but important roles were also played by several carnivores, such as the cat (the animal most often mummified), and then the lion, jackal, wolf, fox, mongoose, leopard, caracal and others. In the temple of Amun Ra at Heliopolis, tame lions were bathed in perfumed water by priests, incense was burned

Egyptian cat, c. 2950 BC, detail from carved panel

for them and they were fed choice pieces of meat to the strains of music. There were decrees to order public mourning at the death of a temple lion, and they were embalmed and entombed (Morenz 1992). A disciple of Buddha rides a tiger to demonstrate his ability to overcome evil, but in Hinduism references to the might of the carnivore are more common than amongst Buddhists. For instance, one of the most important Hindu images is that of Narasimha, the Man-Lion, slayer of demons, saviour of the universe. Shiva, the ascetic sage, was also Master of Beasts and he was associated with many animals including the tiger, deer, elephant and snake (Maxwell 1997).

Nevertheless, it is true to say that in general the major religions of the world are only marginally concerned with carnivorous beasts, and if carnivores are involved at all, our spiritual occupation with them usually focuses on the animals' power and skills rather than on their danger and depredations on livestock. Teaching, warning and protecting people from natural hazards is the domain of literature, of the fable and the story, but it is not of religious concern.

HERALDRY

The human preoccupation with earthly (as distinct from divine) power is much more clearly symbolized by the might of tooth and claw. In heraldry, in the coats of arms and crests of royalty and aristocracy, one finds an expression of the esteem in which different animals are held by people, an expression that used to be as obvious and conspicuous as could be. Heraldry 'relates in symbolic form the hopes and aspirations, the achievements and failures of our ancestors .... Heraldic devices ... were outward and visible symbols of a man's position and influence in society' (Friar 1996). The role of different carnivores in this is quite illuminating, although sometimes surprising.

Heraldry probably derived from the use of lions by early royalty as a symbol of power and courage in battle. Statues of lions were found at the royal gateways of the Hittites, Assyrians and ancient Greeks. Egyptian pharaohs such as Rameses II were accompanied by tame lions, as were the Assyrian kings Assurbanipal and Assurnasirpal, all somewhere around one millennium BC. Lions were used to draw wagons and royal chariots, and the depiction of the Assyrian lion hunt produced terrifying and tragic artistic masterpieces quite unsurpassed in the ancient world. The animals often became playthings, for kings to kill, or to be used as executioners of prisoners. The Romans developed this into a circus, and extended it by having lions, bears, tigers and leopards killed as a public spectacle in the arena. Julius Caesar celebrated the consecration of his government by having 400 lions publicly slaughtered (Guggisberg 1962).

In time, symbols began to replace the real thing, and pictorial representations had to suffice: heraldry evolved. In the early coats of arms only a few different images were used. For instance, in the fourteenth century nine animals were in fashion, most of them carnivores: lion, leopard, pard, stag, boar, dog, horse, bear and dragon. One chose from this list according to personal traits of character or circumstances: for example, the leopard was thought to be a hybrid, a cross between a lion and a pard (but no one was clear exactly what a pard was), therefore it was the chosen heraldic image of someone born illegitimately (Friar 1996).

It hardly needs saying that in the present day the lion still dominates the heraldic scene, and everyone is aware of its royal associations. The lion is formally emblazoned by almost every queen or king. Numerous countries carry the lion and other carnivores in their coats of arms, on government buildings and official stationery. All this despite the fact that wherever these lions (and bears and wolves) occur they are a menace to our society, at least in material terms. They are a menace to be respected rather than just loathed.

Since the earlier days of the use of coats of arms the symbols have diversified considerably, and nowadays many more and different images are in use. In Britain there are some 2000 odd coats of arms described in that classic genealogy of the aristocracy, *Burke's Peerage* (Pine 1956); to see how important different carnivores were, I sampled the first 500 of them. Animals occur in the arms, crest and/or supports of 83% (I excluded fanciful creations such as the unicorn, dragons and gryphons). Of those animals 39% are carnivores, mostly the lion (65% of the carnivores), followed by, in order of prominence, the bear, leopard, wolf, tiger, otter, cat, badger and various others.

Other wild animals are also important, especially the stag and the wild boar, and there are many birds (eagles, falcons and others) as well as domestic animals such as dogs, horses, bulls and cocks. Sometimes such symbols are chosen purely because of name association: for instance the wolves in the arms of the Wolseley family, and the badger (old English 'brock') in the arms of the Brocklebanks. But that does not detract from the symbolic armorial significance of the animals in general.

Heraldic scholars tell us that the lion proclaims majesty, dominance and power. At first sight, the significance of this message is somewhat removed from the lion's menacing role when man meets lion out in the bush, where lions are maneaters and cattle predators. Bears, wolves, leopards, tigers and others have a similar position in coats of arms, and ecologically they, too, are a menace to us. We probably emblazon them up front merely because they put up an admirable fight when we try to kill them.

Use of these animals in heraldry exploits our views of them in our relations with other people. A symbolic leopard or lion on a coat of arms proclaims to enemy and friend that 'I am great, like this lion', or alternatively 'I am the strongest, even this leopard I can subdue'. All the predators and the fantastic beasts on coats of arms, the gryphons, cockatrices and basilisks, were supposed to convey a message; friends would be impressed, ill-wishers terrified by the images of the venomous creatures (Friar 1996).

Apart from serving countries and individuals, heraldry also plays a role in organizations such as societies and universities. Interestingly, in some cases the bearers of the image now contribute to the survival

Heraldic lion

of the original rather than the other way round, e.g. university mascots (such as the tiger of the University of Missouri) spearheading a conservation campaign (Baltz & Ratnaswamy 2000). Finally, one small, lasting symbol of the prestige we accord to carnivores must be the lion stamped on most British and Dutch gold and silver since the sixteenth century, as a standard of its quality. Much of the worldly wealth of previous centuries was expressed in these precious metals, and it was only appropriate that the familiar image of the King of Beasts was its guarantee.

CARNIVORES IN PICTURES

In pictorial art, painting and drawing, wild carnivores have never been particularly popular, certainly not more so than the representation of horses, cattle or deer. Domestic dogs and cats have had their admirers, and many portraits and paintings of landscapes or domestic scenes from the seventeenth century onwards include pets. One even refers to a 'Jan Steen dog', a breed that occurred in almost all paintings of this Dutch master. Special dog and cat portraits were often commissioned in the nineteenth century (Clark 1977). But wild predators in art are scarce, and in general their static beauty is no more admired by artists than that of herbivores. One might reflect that pictorial art generally is not meant to educate, therefore we do not expect painted warning messages of wolves, foxes or bears. One of the possible exceptions is found in cave art, far removed from today's art scene.

The oldest evidence of man's pictorial involvement with wild animals was painted on the walls of caves and rock shelters. Even today this art form is still alive in some places, in Africa and Australia, which greatly helps to understand what motivates the cave artists. In the 1960s I talked to Masai people in Tanzania, who showed me their painting and drawings in rock shelters in the Serengeti (Kruuk 1965).

On the open Serengeti grasslands huge smooth boulders accumulated into a few large rock piles (locally called 'kopjes'), providing wonderful caves and shelters for a host of species. I was in search of my study animals, the hyaenas, and one day when I clambered around in one of these kopjes I found a marvellous frieze of paintings of animals, and people and their arms, in ochres, greys, white and black. The shape and painting of the shields showed that the local Masai tribe was involved, so I went to talk to them.

The Masai showed me how the drawings were made, and explained when and why they made them. Most of the drawings were done during all-male gatherings before raids or expeditions. Whilst talking about their previous exploits, they made drawings and paintings to illustrate their tales – just as a lecturer uses slides. The artists demonstrated how to surround and kill a lion, and which Masai clans were involved (from the patterns on the shields in the paintings), and they also showed exactly where to put a spear if one wanted to kill an elephant. The purpose of these drawings was mostly didactic, and there was nothing magic, religious or supernatural about them, whatever archaeologists may claim for older cave art.

The Masai paintings were especially of cattle, and of people, but there were also many wild animals: lions were particularly common, as were giraffe and elephants. Lions were shown in aggressive postures, facing people with shields, and many a dramatic story accompanied the making of those pictures. In tribal life, Masai men had to demonstrate their prowess by killing lions with spears (an obviously useful ability for herdsmen on the Serengeti plains).

In Masai rock art, lions may have been the most common wild animals, but it was domestic cattle that dominated the frescoes. Elsewhere in cave paintings of wild animals, carnivores were nowhere near as important as in the Masai drawings. In the much older Tanzanian rock art in Kondoa, for instance, Mary Leakey (Leakey 1983) showed that of 474 animal drawings, only 12% were of carnivores, mostly lions and also hyaenas. By far the most common species pictured was the giraffe, followed by eland and elephant. Similarly in the enormous number of famous Bushmen paintings in South Africa, carnivores make up only 5% of the depicted fauna: the artists were much more interested in eland (75% of 1132 animals), although giraffe are absent (Vinnicombe 1976).

These frequency differences must be due to differences in the ecology of the various tribes. Masai are cattle people and pastoralists, defending their flocks against lions, but, from a Bushman hunter's view point, the most desirable quarry is a big wild herbivore. The large African eland is especially good to eat and a single animal can feed a whole village. It appears that these are the animals of a hunter's dreams, and the day-to-day menace of predators or competitors is less interesting.

I found virtually no carnivores in the prehistoric cave paintings of France and Spain, and it appears that the Sahara cave art is similarly deficient (Lhote 1958). The conclusion can only be that cave art is largely concerned with food animals, with people's preferred prey, with humans themselves, and in later times with livestock.

Going from caves to more recent illustrations of animals, one finds that in early medieval European art, presentations of carnivores were often symbolic rather than realistic. One sees monsters tearing victims apart, and representing evil on earth. Sculptures on Romanesque churches depicted lions, wolves and foxes as well as other and imaginary animals, and there were links with the Reynard literature of the day. Where animals were shown closest to nature, at least as in those days one imagined nature to be, was in the bestiaries, which I have already mentioned, beautifully illuminated manuscripts to be found in several medieval centres of learning (for instance there is a famous one in Cambridge, UK). The role of the bestiaries was clearly encyclopaedic, although both the descriptions and the magnificent illustrations were often literally fantastic, with a now dream-like quality. However, as T.H. Huxley wrote, 'Ancient traditions, when tested by the severe processes of modern investigation, commonly enough fade away into mere dreams: but it is singular how often the dream turns out to have been a half-waking one, presaging a reality.' (Clark 1977.)

In the twelfth and thirteenth centuries when the bestiaries were produced, few people had any conception of animals outside their own surroundings, and the artist's imagination was left a free rein. Carnivores were pictured prominently, but when one considers the role of these animals in art one has to concede that the illustrations of the bestiaries only serve as a corollary to the text, rather than as art for its own sake. However, throughout the following centuries several artists made a special effort to portray animals. Lion and leopard were painted by Giovannino de' Grassi in the fourteenth century, and after that the fame of painters such as Albrecht Dürer (fifteenth century) was partly based on his fascinating animals, studies which included a wonderfully life-like resting lion. The genius Leonardo da Vinci, also living in the fifteenth century, led an urban life without exposure to wild animals – a pity, because his drawings of domestic cats, dogs and a captive bear have great character and are beautifully realistic. In those days animal art found its subjects in the rare collections of captive animals (or in pets and livestock), but despite that limitation the drawings of lions by a master like Rembrandt are unusually powerful.

Only much later, in the eighteenth and early nineteenth century, did some of the Romantic painters turn their attention away from captive wild creatures, and people and domestic animals amid surrounding landscapes, and start to look towards the wilderness. In the spirit of that time it was William Blake who wrote 'The tigers of wrath are wiser than the horses of instruction' (Clark 1977), implying that raw natural instincts are better guides than man-made rules. It was the opening of the artists' minds to animals in their natural environment. For instance a Frenchman, Antoine-Louis Barye, was a sculptor and painter whose paintings of tigers show what was for those days an unusual truth to nature. He depicts his wild animals walking quietly or lying down in nature, rather than killing or being killed.

Barye's more famous contemporary, Eugène Delacroix, painted tigers, lions and leopards, sometimes captivated by the languor of his subjects against thunderous clouds, but more often depicting fierce battles of predators against horses and man. His view was that 'Art does not consist in copying nature, but in recreating it, and this applies particularly to the representation of animals' (Clark 1977). His paintings of predatory violence were not very popular at the time, and people preferred his more gentle horse scenes. Henri Rousseau painted jungle scenes in the nineteenth century, in which animals were central, and carnivores sometimes featured.

Clark wrote that public opinion against killer animals in paintings shows that we do not want too much violence, at least not in art. It 'confirms the view that all is well as long as strength is controlled by skill', and if the predator wins we do not want to be reminded of it. According to him, artists leave dangers and nuisances out of the picture, because they do not please, and painting is not concerned with didactics (although that does not appear to hold when artists paint scenes of war or violence in religion). I think that the rarity of carnivores in art suggests that either the artist does not perceive them as any more beautiful than other animals, or that the opportunity to paint them does not often present itself to painters other than the modern wildlife artist. But in contradiction to the above, along comes Salvador Dali, using the image of the most fearful of maneaters opposite the most vulnerable of people, with two wonderful attacking tigers and a nude, sleeping girl in his dream picture. If ever there was a scene of threat, this is it.

Our culture is suffused by carnivores of one kind or another, sometimes as villains, sometimes in roles of magnificence, demanding respect for their strength or beauty. There is little doubt in my mind that at its most basic our art, literature, heraldry and many of their derivatives have a role of instruction (amongst others), and as such they are an important extension of the learning part of our anti-predator behaviour. In a purely biological sense, through art and especially through literature, we can send alarm signals to our conspecifics.

These alarm signals have been embellished and magnified, and they have acquired their own significance and started their own life, but in their roots the original function can still be discerned, and it still serves to warn us.

Foxes and dustbins

12

The future
Effects of humans on carnivores: urbanization and survival

Late one evening, in darkness, I was wandering through the narrow streets of Harar. It was a scene straight from the Middle Ages, with old houses leaning over the streets in the ancient walled city in the lowlands of Ethiopia. There were high gates in the city wall, which was still virtually intact, although crumbled in places. The heat of the day had lessened somewhat, but not much, and the moon dominated. The city was quiet after the roaring daytime activity, the markets, the traffic around the outside of the wall, and the throngs of people. Now at night the bark of a dog stood out clearly, and people had retired to their houses. Not all of them, for one or two slept in doorways, oblivious of my passing.

Just ahead a familiar shape crossed, its shambling gait better known to me as that of a hunter from the Serengeti plains. The large hyaena disappeared through a gap in the wall, having walked within a few feet of a sleeping woman. Minutes later two more withdrew around a corner. Here in Harar, the spotted hyaena was a well-known, tolerated scavenger in the town, clearing up the bits and pieces, never leaving a single piece of bone or meat. Just outside the city walls I found a local man sitting on the ground, wide awake despite the hour. He was surrounded by a dozen hyaenas who were interested in titbits. The man knew the individual hyaenas by name, and he fed them bones by hand and even from his mouth.

My visit to Harar happened 30 years ago, but things have not changed. The hyaenas still patrol the city, cleaning the streets, and a benefactor still feeds them by hand, at night. The spotted hyaena, maneater when it suits it, has been urbanized in Harar.

Similar scenes, different settings, a BBC TV camera focused on a pair of mating foxes: the picture was wonderful, close-up, the two animals with only one thing on their minds. The camera shifted, a hedge came into view, then people hurrying along with their brief-cases, and in their cars. The rich tapestry for the foxes' love life was a small, very urban garden, and busy city dwellers were the backdrop to this little jewel of natural history. Foxes are common in British towns such as London, Bristol, Oxford, Birmingham and others, even right in their centres. Scavenging is their source of livelihood, and it is a prolific source. One evening, walking our dog just outside Oxford, I counted 23 different foxes, including cubs, in a span of 2 hours; it showed me that densities of these animals in and around towns reach surprisingly high figures, far higher than anything in the wilds of nature.

Not so long ago, in the early 1990s, the BBC made another film on wildlife in British cities, which I vividly remember. The scene was a living room in Birmingham suburbia opening out on a garden, it was evening, the television was on and an elderly man was watching it. A badger came wandering into the room from the garden, obviously confident and familiar with the place. The man threw some peanuts in front of the television set, and the badger happily ate away, meandering about in front of the newsreader. Then another badger joined, and another, and more peanuts were offered. At the end of the shot there were over 20 badgers in the room, a heaving dark-grey mass with peanuts as its main objective.

Urban wildlife has become part of the town scenery in many countries. Carnivores play a major role in it, probably because their abilities to scavenge for a living predestines them for this niche. Just to show the extent of the changes, I will mention some of the major participants in this game of commensals.

The most common urban and suburban carnivore species in Europe are the red fox (Macdonald 1987), Eurasian badger (Neal & Cheeseman 1996) and beech marten (Broekhuizen 1983), but several others also use the human habitat. We find the otter, polecat, and the introduced American mink in and around villages, perhaps not scavenging as much as the first three species mentioned, but exploiting the rich resources often created by humans. In Romania, even the wolf has become a city dweller, to be seen under the street lights at night, roaming the outskirts of town and visiting garbage tips (Promberger 1996). Similarly, brown bears have also taken to life in towns there.

In North America the raccoon is a well-known urbanite, with many cities supporting large populations. Red foxes there are also attracted to the fleshpots of towns, as are coyotes and skunks. More spectacularly, brown, black and polar bears take to garbage in a big way, and their presence around suburban houses, camps, or even inside towns, may cause nasty accidents. The main claim to fame of the Canadian town of Churchill, Manitoba, on Hudson Bay, is the presence of numbers of polar bears. Anywhere in the American or Eurasian Arctic a village can be the haunt of arctic foxes in winter.

Striped hyaenas hang around human habitation wherever they occur, throughout the Middle East and the northern parts of Africa. Even in medieval bestiaries they are shown as grave robbers, and the palaeontologist Anthony Sutcliffe found their dens in Kenya strewn with human bones (Mills & Hofer 1998). Hyaenas are not alone in this gruesome specialization, as wolves are also known to dig up human corpses around towns and villages. The two species often occur jointly around desert settlements in the Middle East, and I have watched them scavenging together near kibbutzes in Israel (Kruuk 1976b).

In Africa, the genet, wild cat, spotted and striped hyaena, Egyptian and black-tipped mongoose, honey badger, three species of jackals and others are all keenly drawn to the magnet of villages and towns, although they are often put off by dogs. As in other continents, in Africa livestock presents an attraction on its own, providing food for a range of predators.

Some of the commensals to human communities have been there for many centuries, their role firmly established and largely accepted by people. But most have only come into this position very recently. The invasion of British cities by foxes and of European continental towns by beech martens took place during the second half of the twentieth century (Harris 1986; Kugelschafter *et al.* 1993), as did many of the other examples that I have mentioned.

There can be no doubt, therefore, that some carnivores benefit greatly from the presence of people, increasingly so, and that humans could be said to play a positive, although passive, role in their survival. But this may go further, with people actively transporting predators into new countries and new habitats. An entire continent, Australia, would be without carnivores were it not for us, and we can boast the same for scores of islands large and small. Some of the details and species involved have been discussed in Chapter 8.

Much of this animal dependence on human society is still evolving. The recent developments suggest that there is a trend under way,

Polecat

and that more carnivores will follow our society in its present expansion, using a new and promising habitat. Human populations are increasing rapidly, and a greater proportion of us is living in cities. The urban carnivore has probably found a lifestyle for the future, which is more than can be said for many, many others amongst their relatives.

CARNIVORES UNDER THREAT

Many natural habitats are disappearing, and many prey species are declining. Amongst the predators, one would expect a priori that the populations of narrow specialists are most at risk from such changes, i.e. those species that may be confined to a small range of kinds of prey or of habitats, in contrast to the opportunist urban foxes and raccoons. But other, less specialized species also have serious problems now because of the impact of *Homo sapiens*, and they will face much more serious, if not terminal, problems in the future.

In November 1998, the London Zoological Society hosted a meeting with the evocative title 'Has the panda had its day?'. It reported on a wealth of new research on conservation strategies and species survival, but it also noted with concern that conservation in general focuses on the survival of large, charismatic animals with the panda as flagship. Much more money has been spent on their conservation than on species with a low profile, or even on entire habitats. One thinks of the tiger, which joined the panda in receiving special care as a show species. But not only is the survival of even those species in doubt despite this special attention, in addition, and arguably because of this policy, the conservation of other, less conspicuous species has suffered considerably. Some scientists just take it as unavoidable that we will lose much of our fauna, e.g. Joel Berger summarizes a recent paper with 'The earth is likely to lose large carnivores in the near future' (Berger 1998), and suggests that this will have important negative consequences for wild prey species.

All this sounds gloomy, and it is easy to decry conservationist merchants of doom. For instance, some extinctions may be natural, as man is not the only influence on survival. After all, there have been many extinctions before now, before humans made their major impact on the world. As I explained earlier, the balance of evidence suggests that, for those earlier demises, people were not the main cause: there were other dramatic changes in the environment, and for at least some of the extinction waves we have an alibi as they happened before man's arrival. We cannot be certain, but the chances are that we need not have a guilty conscience about all the disappearances, even during the early days of our own evolutionary history: species come, species go. Moreover, the optimists can argue that at present we have a wonderful range of national parks and reserves right around the globe, and many individuals of almost all species are well protected. In fact, it seems to be well-nigh impossible to eradicate some of the carnivores that some people want to be rid of in some places, such as foxes and coyotes and many others, and populations of many species are booming in our immediate presence.

Such arguments of optimism ignore the reality of the great divide between rich and poor, between developed and developing countries. In a few rich parts of the world it seems likely that we will be able to maintain what there is, to keep more or less intact at least a number of those wonderful ecosystems in which carnivores occupy such a conspicuous place at the top. This is largely the position in North America, in Britain and in many other countries in Europe. Of course, the future here is by no means secure and many aspects of wilderness as we know it will disappear. But with enough effort there is a reasonable probability that we can maintain most of the species that we now have left, in their own, more or less natural habitats.

In these richer countries, people are prepared to pay for the knowledge that their country still is the home of species like the wolf, or the puma, or the badger, the wildcat and the otter. As an example, in Britain ecological economists have demonstrated an average willingness of the average citizen to pay, as a one-off tax contribution, £11.91 ($19) just for the conservation of otters (and a bit less for some other species) (White *et al.* 1997). Carnivores especially have benefited from this public sympathy, and there have been costly and successful reintroduction schemes in many areas of Europe, North America and South Africa. These schemes involved lynx, otters, badger, and brown bear amongst others, and in South Africa all the large cats and wild dogs have been rehabilitated in parts of their former range.

However, in the poorer developing countries the issues are different. Where the future of ecosystems and wild animals is pitted against the survival of people's families, then wilderness has little chance. I will mention one example, which has shocked me because it concerns an area in which I used to be personally involved, a sublime wilderness that I see as the Eden of carnivore diversity.

The Serengeti is nature's prime World Heritage Site. Situated mostly in north-western Tanzania, but with Kenya's Mara a part of it, the Serengeti ecosystem is an area of some 25 000 km^2, approximately the size of New Hampshire or Vermont in the USA, or of Wales in Britain. Amongst the vast number of animals there are more than 25 species of carnivores, and many hundreds of thousands of ungulates move around in a continuous migration. It is a fabulous wealth, and the spectacle has given millions of visitors the experience of a lifetime. But that world is now eroding fast.

In both Kenya and Tanzania agriculture is moving right up against the borders of the national park, which makes up less than half of the entire ecosystem that contains the animal migrations. Whereas in the 1960s one could drive all of the 300 km from the Serengeti to the nearest large town with frequent sightings of herds of wild animals, a sea of cattle and wheat is now pressing against its boundaries. This is not surprising: numbers of people have increased fast, as have their demands for food and wealth. Over many years now, the human population around the national park has increased by an average of about 2.8% per year (Campbell & Hofer 1995).

The income that local people in these areas derive from the national park and from tourism is pitiful, because most of the tourist revenues, including the high entry fees, go to the countries' capitals, or to the tour companies, or they stay in the national park. Clearly, this could be remedied, but the fact remains that even if a justifiable part of the revenue were to go to the communities of local people, it would still be more lucrative for them to grow wheat in the Serengeti than to run it as a wildlife reserve.

The different degrees of benefit under various management regimes have been calculated in detail for the Kenyan region of Mara by the wildlife economist, Mike Norton-Griffiths, and these figures are probably valid for the entire Serengeti (Norton-Griffiths 1996). He showed that in 1994 in the Mara the Masai landowners earned annually per hectare the equivalent of US$0.35 for areas used by wildlife (from tourism), $1.99 for ranches used for livestock and $6.25 for land under agriculture (cereals). The potential for agriculture on the rich volcanic soils of the wildlife areas (the reserves) is excellent. The discrepancies in income as shown by Norton-Griffiths explain exactly why large parts of that vast region supported thundering herds of wildebeest in the 1960s, and huge, undulating wheat fields at the turn of the millennium.

The management options to rescue biodiversity under such conditions are limited. Consumers of the joys of biodiversity need to pay the guardians, i.e. the local communities in the developing countries, to forego the advantages of developing their lands. These 'opportunity costs' are high, and it is doubtful that the relatively large amounts of funds involved can continue to flow from the rich countries year in, year out, especially since the benefits for the rich countries are not tangible. An organization such as UNESCO might take a lead, but it is difficult to be optimistic.

The Serengeti is not alone here. The argument that people need to scrape the maximum income from their own soil applies all over Africa, and in many of the poorer countries in Asia and Latin America. In developed countries we may ignore such economic considerations against the preservation of biodiversity, because with a basic lifestyle secured we can afford an interest in art, wildlife and environment and we are prepared to pay for it. But this does not apply when children are starving on the fertile grasslands of Africa: there, local communities need the best possible economic return on their lands and that is through agriculture. For them, the preservation of biodiversity is an unaffordable luxury.

The above arguments are concerned with habitats, and with areas, with whole ecosystems in which carnivores play an important role. But similar considerations apply to the utilization of the individual species themselves. The strongest incentives to obtain maximum benefit of animal products occur in the developing world. There, 'maximum benefit' often does not mean 'biodiversity conservation' or tourism or some other sustainable use. It means human consumption of the usual prey of the carnivores (often by poaching), it means maximum pressure to obtain meat and other products such as skins from the animals, maximum incentive to remove any animal that may be harmful to livestock or to animals caught in snares or traps. The policies of hunger are from dire necessity egoistic, non-ethical and short-sighted.

In relatively few situations, sustained use of carnivores by tourism can provide substantial income. Tourism places values on them, which are translated into revenue. For the Amboseli National Park in Kenya it has been estimated that a single lion is worth US$27 000 per year (Western & Henry 1979). In that case, in that one area, the conservation of lions is easy to justify, because tourism may well be the most lucrative form of land use. In Amboseli agriculture would be difficult; but in many other places the balance is weighted heavily in favour of agricultural exploitation.

Many of these comments could apply to conservation in general, and to animals of all kinds. But carnivores are more vulnerable than most other animals, and they face a set of very important additional handicaps. Firstly and most obviously, carnivores need prey. Not only are they subject to perils directed immediately at them, but they are also affected by any threat to levels lower in the food pyramid, by anything that lowers the numbers of their prey species, because predator populations tend to be food limited (Chapter 9).

Secondly, carnivores tend to have large home ranges, some of them covering hundreds of square kilometres. If suitable habitat is to be found only in small reserves, such animals have little chance of safe havens where they are protected against deliberate assaults by people as well as against traffic and other hazards. In fact, the mean size of the species' home range is a good indicator of its vulnerability to extinction (Woodroffe & Ginsberg 2000).

A third reason why I believe that carnivores are probably extra vulnerable is disease. Many potentially lethal diseases can be transmitted from one carnivore species to another. Burgeoning human populations bring with them massive numbers of dogs and cats, and especially in the poorer areas of the world these domestic animals carry viruses such as distemper, rabies, cat flu and others. Contact between domestic and wild carnivores is often fatal for the latter (Woodroffe *et al.* 1997).

Fourthly, as predators, some of the carnivores attack livestock and people, and for that reason alone, they are persecuted (see Chapters 4 and 5). Fifthly, predators are more prone to accumulate pollutants than are herbivores, because being at the top of the food pyramid, they have that extra level of accumulation. By eating a fish or an earthworm, an otter or a badger ingests the pollutants that have already been accumulated by the prey. A carnivore concentrates the herbivore concentrates.

Furthermore, carnivores have many attributes that make them useful to people: furs, and supposed medical and magic properties. A carcass may be worth a great deal of money, and if you are poor, carnivores are worth killing for any of those reasons. Hence the predicament of the tiger. Lastly, numbers of predators are inherently lower than numbers of their prey species. This means that the process of population fragmentation, inevitable in the wake of land development by people, leaves carnivore populations smaller than those of other, comparable mammals; as any statistician can demonstrate, small populations are risky, and subject to extinction by chance disasters.

Despite all this the animals are still there, wild carnivores are still with us, and what all these special vulnerabilities suggest is that with their many handicaps, the average carnivore species must be extremely resilient. On the scale of endangered animals they do not appear to fare worse than the average mammal and their risk of extinction is

The red (or lesser) panda - an endangered species

not significantly different (Mace 1995; Purvis *et al.* 2000a, 2000b). As another example, the proportion of species predicted to be extinct in 100 years is 16% for canids, whereas for vertebrates as a whole this is 15%. But some carnivores are in dire straights, and some of the very charismatic ones are at the very risky end of the spectrum of extinction potentials.

Recent tales of woe are ample illustrations of typical troubles faced by carnivores. Take the African wild dog, of which there are now only a meagre 3000–5000 left in the world. Only a few decades ago there must have been many tens of thousands, although nobody counted them then. Throughout their sub-Saharan range they have been shot, snared and poisoned because of their predation on livestock, a persecution that was relatively easy because the animals are diurnal, live in large and dense packs and are not at all shy of people. In one country alone, Zambia, the government vermin control units killed about 5000 wild dogs between 1945 and 1959 (Woodroffe *et al.* 1997), and the managers of many national parks and reserves anywhere in Africa shot wild dogs on sight 'to give antelope opportunity to develop optimal numbers' (Attwell 1958), as was recorded in a conservation journal even as recently as the 1980s. People and wild dogs did not mix well then, and they still do not now.

The wild dog populations that are left are very fragmented. There are few areas large enough to accommodate their enormous home ranges (the average for a female is 823 km^2), because everywhere in Africa there is a relentless increase in numbers of people. Even in the Kruger National Park, at 22 000 km^2 about the size of Israel, about 47% of all mortality amongst wild dogs is caused by people outside its boundaries (Woodroffe & Ginsberg 2000); and even in such larger areas, the dogs are now subject to several diseases which they probably pick up from domestic dogs, especially rabies and canine distemper. For instance, the Serengeti lost its last few wild dogs in the 1990s because of these two diseases (Burrows 1995), although the underlying population decline may have been due to competition with other carnivores. A few countries still have reasonable numbers of wild dogs in a few places, such as South Africa (in the Kruger park), Botswana, Zimbabwe, Zambia and southern Tanzania. But the decline continues, and there appears to be little hope for their long-term future.

The Ethiopian wolf (formerly called the Simien jackal) is the most endangered member of the dog family in the world. There are only a few hundred of them left, surviving in small, fertile patches high in the alpine zone of the mountains of Ethiopia. They are beautiful, brown or orange-red animals, with white throats and underparts, about the size of coyotes. The prognosis for their long-term survival is far from good (Sillero-Zubiri & Macdonald 1997). The main problem that the species faces is the disappearance of their habitat of wild grasslands on the mountain slopes, where they catch their main prey such as giant mole rats and grass rats. The rapid decline of this habitat is due to agriculture and overgrazing by livestock. As a result, Ethiopian wolf populations are now very isolated and becoming more so, with the consequent increase of contact with people and their domestic dogs. Poaching and diseases kill many of the animals, and the process is accelerating.

Humanity has looked in awe at the lion for more than three millennia, and people's fantasies, aspirations, admirations and fears are locked up in its image more than in that of any other carnivore. But it is without any doubt the lion's closest relative, the tiger, which is the carnivore par excellence. It is the animal of solitude, of proverbial beauty, majesty, strength and agility, and with its magnificent camouflage it disappears into its jungle background as a killer. It is responsible for many human deaths and it is a scourge on livestock. It is also an animal that is now close to extinction in the wild.

At the beginning of the year 2000, the total number of tigers left in the vast forests of South-East Asia, the area between India, the Far East of Russia, Malaysia and Indonesia, has been reliably estimated as no more than 5000–7000 (Nowell & Jackson 1998; Seidensticker et al. 1999). There were eight subspecies, of which three have become extinct in the last couple of decades and five still survive. However, subspecific status means little apart from occupation of a given area, because the variation (in size, colour, skull measurements) between subspecies is no greater than the variation within (Seidensticker et al. 1999); they are all in the same boat.

Tigers used to live in an area that covered 70 degrees of latitude and 100 degrees of longitude, from the Caspian Sea to the Pacific, from Siberia down to Java and Bali. Now, a tiger map of that area shows only a few small black patches. Of course, we all knew that tigers were vigorously persecuted over the centuries, but they survived in respectable numbers. In the recent past, however, two distinct crises have hit the species.

In the 1960s and 1970s, the unprecedented rapid loss of habitat, and over-hunting for sport and for skins, was combined with the need to solve conflicts with the human population. Royalty had also shot tigers by the score. All this drove numbers down to dangerous levels, although it took some time before it was realized that harm was being done to populations. But then alarms were raised, with a great deal of publicity. Governments initiated action, and the world began to frown upon the antics of royalty and others that endangered this wonderful symbol of wilderness. Several official policies to protect the tiger and to assuage the alarmed conservationists were successful, and there was a general feeling that the crisis was over. Then, in the 1970s, tigers seemed to be safe, albeit in much smaller numbers, and the shooting had stopped.

However, in the 1980s and 1990s a second crisis hit, and its effects were worse than those of the first one. The acute economic problems of tens of millions of people, and the easy availability of firearms, triggered a rash of poaching, targeting the tiger and its bones for the trade in traditional medicines. I have mentioned the medicinal uses of tiger parts in Chapter 7; they are many and diverse. Traditional medicine pays handsomely, and more than one fifth of the world's population is prepared to use it. The wealth of Hong Kong, Taiwan, China, Japan and Korea is readily available to the man with the gun in the forest, as long as he produces tiger bones.

A conservation campaign against tiger poaching started in 1993, just after a huge haul of 500 kg of tiger bones was seized in India. This was recognized for exactly what it was: evidence of serious and massive poaching of the species everywhere in its range. Even a relatively small country like South Korea imported 8951 kg of tiger bones between 1970 and 1993. Governments agreed to clamp down on the trade, and today only Japan has not banned it. But there is considerable residual and illegal activity, and the verdict of an impressive book by the world's tiger expert John Seidensticker is that 'the tiger probably cannot survive the pressure of even this residual trade for long' (Seidensticker *et al.* 1999). He is not even talking any more about the fast erosion of the tiger's forest habitat, which is serious enough on its own: this has now become a secondary hazard to the species.

A possible solution that has been mooted is tiger farming, providing products legally and perhaps more cheaply (Schaller 1996); however, opponents point out that this would stimulate demand, and would make policing more difficult, because illegal possession would be almost impossible to prove. Also, the provision of alternative products has been considered, such as fake 'tiger bones', and bones of other species. That also does not seem to be a likely solution, as the rich devotees of traditional medicine would still want the best, i.e. the animal itself. The tiger tale is not likely to have a happy ending.

The giant panda is, in effigy, amongst the best-known animals in the world. As logo of the World Wide Fund for Nature it has become the symbol of conservation, and the trade in panda cuddly toys is a trade in millions. I am not sure that all this is a good thing. The panda image has become something of a joke, and its curious appearance does not help: it may appeal to lovers of teddy bears, but is it still an effective cry for help for our fauna? Is it even an effective cry for help for itself? At the moment, on any stock market its shares would be very low.

There are now somewhere around 1000 giant pandas left, in small isolated pockets scattered over a wide area (Schaller *et al.* 1985). Many of these populations have fewer than 50 animals. Even when taking all pandas together, 1000 is only a small number in population terms, and there are several major, combined threats to the animals. The first and overriding problem is that the pandas' habitat is being swallowed up by advancing human populations, so the animals become confined to small areas. This would matter less if the pandas were not so extremely specialized and dependent on bamboo, which tends to grow in large single-species stands. Bamboo dies off after flowering, and major

Giant panda

mass flowerings plus die-offs may occur once every 20–120 years, depending on the species. A major die-off of bamboo leaves many pandas to starve to death, as they do not have the option any more of moving elsewhere.

The rarity, popularity and appearance of pandas makes them an attractive object for poaching. A panda skin is worth over $10 000, which is a fortune in China; owning one is a status symbol in the richer countries of Asia. But the risks have increased as well, and panda poaching now carries the death penalty in China. Nevertheless poaching continues, and the odds are stacked heavily against the species in the wild. One obvious but temporary solution, i.e. captive breeding, seems to be extremely difficult, despite the many competent attempts that have been made. At present various new techniques of *in vitro* fertilization appear to be promising.

Almost at the other end of the carnivore size spectrum from the panda is the black-footed ferret, denizen of the USA, which became extinct in the wild in the 1980s. The species was nearly wiped out because the prairie farmers poisoned and shot almost its only food, the large rodents called prairie dogs. It became virtually extinct in the 1970s, but then one last colony was found in Meeteetse, Wyoming, in 1981. This was intensively studied, and the researchers noted a sharp decline, from 86 animals in 1982 to 15 in 1987. It was decided to take six animals in captivity to breed, but all six died of canine distemper within a week. In the wild three more died and only six remained, which were all then caught and vaccinated. No more deaths occurred, and a captive propagation project began (Reading & Clark 1996). The captive breeding was very successful, to the extent that in 1991 the first black-footed ferrets were reintroduced into the wild. Since then, captive ferrets have been released at several sites in Wyoming, Montana, South Dakota and Arizona. The species is back again with hundreds of its members in the wilds of America. It may have gone through a genetic bottleneck, but it was saved, by the skin of its teeth.

Hopes are that a mustelid species of similar size to the black-footed ferret may yet be rescued in Europe: the European mink. Its long-term, slow decline suddenly accelerated when the American mink was released in Europe in the 1920s, and the European mink rapidly disappeared from many countries (Youngman 1982; Maran & Henttonen 1995). It now still occurs in Spain, and in a last few East European countries where it is declining fast, such as Russia, Belarus and Romania. There is little chance of long-term survival on the continent in the face of the rapid increase of the aggressive American species, and of heavy trapping pressure to supply the local fur trade.

A project is now under way to establish a population on one of the larger Estonian islands in the Baltic sea, Hiiumaa, by releasing captive-bred European mink. The American mink on the island, remnants of the stock of a mink farm, have been removed, and the first few dozen European mink were reintroduced in 2000 and 2001. The prospects for an established population are good, but even if the project succeeds, the result will be only a very small safe haven for the species, a small black dot on a map that once covered almost the whole of Europe.

A much more cheery success story is that of the sea otter (Riedman & Estes 1987). Its natural range extended from Japan along the northern seaboard of the Pacific Ocean as far as Mexico. Russian explorers discovered the Aleutian Islands and Alaska in the 1740s, and brought back tales of large numbers of sea otters. They told of the ease with which the animals could be taken, and of their excellent fur, and soon the slaughter was under way, opening up a vast fur trade with China worth millions of dollars. By the time that the USA bought Alaska from Russia in 1867 (especially for the fur trade) there were so few sea otters left that it changed ownership for a mere seven million dollars. Exploitation continued unabated under American authority, and when finally a total protection of the species was declared, sea otters were on the brink of extinction. Fewer than a 1000 were left, scattered over 11 locations; many people thought they could not survive.

It took a few more years for illegal hunting of sea otters to stop, and they died out in several of their 11 known areas. Then the species began to recover. The first populations increased unaided, but from the 1960s onwards numbers were translocated from the sea otters' strongholds to areas where they had not yet returned. The animals are common now along many of the coasts where they had disappeared (Ralls *et al.* 1996), and the world population now stands at several tens of thousands. Interestingly, in the temporary absence of the sea otters new shell fisheries had built up along many of the coasts, especially for abalones. When the otters returned they made serious inroads on shellfish stocks, often putting fishing people out of business. There were vociferous complaints and demands for sea otter control, demands that until now have been successfully resisted by the conservation authorities.

Sea otters are back along the Pacific coasts of Russia, and in North America from Alaska to California. Everyone hopes that they are back to stay, giving thousands of people the chance to see them. Yet there still are concerns: in the last few years of the last millennium populations in the USA were declining again by some 4% annually, for reasons as yet unknown. The sea-otter expert Jim Estes observed increased predation on them by killer whales (Estes *et al.* 1998), there are reports of Russian sea otters being heavily poached, and there is always the worry about potential oil spills.

Some you win, some you lose. Some species or areas are saved and protected, others fall. The problem, of course, is that in conservation in the longer term, every winner can yet be turned into a loser, but no loser can ever be made a winner again. The carnivores that can turn themselves into commensals on human society will do well. That means the red foxes, raccoons, Eurasian badgers and stone martens of this world, and we may expect several more to climb onto that bandwagon in the next decades.

Furthermore, some of the species living in man-made habitats appear to be totally indestructible; people could not exterminate them even if they wanted to. Species that exist in habitats too extreme for our aspirations should also be quite safe, such as those living in arctic climates and deserts. Many others will escape capture, eradication or the destruction of their habitats, by one means or another. However, as I discussed earlier, some species, including many of the larger ones such as the tiger, panda, several other bears, African wild dog, Ethiopian wolf and others, are now seriously endangered. Being highly specialized in their feeding habits, or attacking livestock, or having very large home ranges, are all characteristics that diminish chances of survival for those predators, and the scale and speed at which the world's habitats and populations are changing makes one fear that there will be many more extinctions soon.

In developing countries, where often people are up against the threat of starvation, those carnivores that compete with humans in one way or other are endangered, and those animals will not be protected unless they pay their way, unless they create wealth. In theory, richer countries could help in this process by paying the fare for threatened fauna in the developing world. To some extent this happens through tourism, or by direct conservation aid from richer countries or international organizations, or in dedicated schemes such as the Darwin Initiative of the British government.

But such a basis for continued existence of species or ecosystems is a very tenuous one, riddled with long-term uncertainties. Tourism is extremely sensitive to political instability, a problem that is relevant especially in developing countries. Foreign aid tends to be short term, and in any case people in the developed world, in the countries providing the aid, are likely to be sceptical about long-term support for an environment from which they will derive no benefits, and which probably they will never see; and if that scepticism is not enough, effective support for the environment is much more expensive than, for instance, support for the arts.

Predators may have to go from developing countries where people will need to exploit the habitats that were formerly available to these animals, where man has to exploit these areas for long-term, maximum profit, and not for the interests of carnivores. On the positive side, developing economies are able to sell carnivore products, such as furs, to the richer world and this could maintain populations on a sustainable yield basis. But that horizon has some clouds, too, because of the strong animal rights movement in the richer countries of the world against animal exploitation, against the use of furs, and against hunting. These are the very kinds of exploitation that if properly controlled and regulated (unlike the present-day poaching of, say, the tiger), could save species.

Because many habitats are changing rapidly and very drastically, amongst them various types of forests and savannahs, the threats to survival of species that are specially adapted to such landscapes will increase, adding further hazards to the many others in the tenuous relationship between people and carnivores. By projecting forwards the trends in Africa, South America and Asia of the last 30 years, there can be no doubt at all that over, say, the next 100 years a considerable number of carnivore populations, and probably whole species, are likely to disappear from the wild.

There have been very good times for carnivorous animals in the early millennia of their evolution, but a declining trend in their diversity started even before people arrived on the scene. Firstly there were the many predator species in the order Creodonta, as well as the marsupial carnivores. They blossomed, and then disappeared again. Then the Carnivora proper arrived, and these predators and competitors may well have had a firm hand in the demise of their predecessors the creodonts and carnivorous marsupials; they occupied the empty niches left by the creodonts. The Carnivora had a heyday, and the number of species increased, but then, slowly, they themselves also began to dwindle. In the near future the trickle of their departures may burst into a flood. This time there is no other order of mammalian predators involved, ready to take over – only people.

Despite all such gloomy predictions, here and there we see glimmers. Apart from the return of the sea otter, North America is also seeing one of the martens, the fisher, recovering in its forests after trapping stopped, and wolves are being restored to wilderness where formerly they had been shot and poisoned out, to areas throughout the USA, including the southern states (Siminski 1998; Mech 2000). Pumas appear to be expanding back into many places again, and in 1990 in California (which holds a large number of pumas) a total of 30 million US dollars was allocated for the next 30 years to provide habitat for the puma and other threatened species (IUCN 1996). Several countries in Latin America are determined to maintain well-protected areas with a full complement of large predators in them, countries such as Argentina and Belize. In Scandinavia, the wolf, brown bear, lynx and wolverine are doing well again, to the extent of causing problems. Lynxes are being introduced into many central European countries, and wolves have returned to Switzerland and France without human help. Tigers are reported to be doing well in areas of Nepal, with a recent increase of 50 tigers to about 300 in the country (Planetsave.com 2001).

We may yet live in hope. We can remember that carnivores have a remarkable resilience, and that so far they have done no worse in the face of the human onslaught than the average mammal has, despite their many handicaps. The wiley fox Reynard and his family may yet survive, again.

One special feature that may keep our hopes alive for these animals is the strong, behavioural response to predators, a response that is characteristic for probably all birds and mammals, the non-aggressive attraction to them that we have termed curiosity (see Chapter 10). As in those other mammals and birds, curiosity is part of our defence mechanism, enabling us to learn about our foes. But in our species and in our culture it has gone further than that: through curiosity, we not only learn how and when to be on our qui vive, but also we see in those carnivores other hunters, and in our imagination we get involved in the stalk or chase, and we admire.

Whatever the exact biological mechanism of this behaviour, there is no doubt that there is a deeply engrained appeal of the fleet-footed or stealthy jaws, despite or perhaps just because of the threats they pose to us. As a fortunate extra, the love of our pets, especially in western society, will always rub off on their wild relatives, and make us see them as individuals. One must hope that these special appeals of carnivores are strong enough to allow for the right decisions to be made on the survival of the animals.

Cheetah

Epilogue
Changing views of carnivores, individuality and conservation

Now, at the turn of the millennium we are seeing animals, especially carnivores, from a perspective that is very different from the one we had before. Our views are changing, and this is a process that is going remarkably fast. Let me give an example.

In the late 1960s the magazine *Life* featured a set of spectacular photographs of a leopard catching and killing a baboon, in Africa. The pictures showed every detail of the sinewy predator in the African bush, the blurred, lightning-speed action, and the intensely frightened expression on the face of the baboon at the final moment. The photographs were stunning, almost literally so because they were so excellent that one identified totally with the victim. Later, it became known that the entire scene in those photographs was set up. The leopard was a tame animal, the pictures were taken on a ranch in Kenya, six baboons were bought from an enterprise that provided primates for medical experiments. The photographer sat in a vehicle that was being driven alongside the leopard, and the baboons were thrown in front of the leopard from the vehicle.

Such deliberate and horrible cruelty for the sake of a picture is unlikely to happen today, only 30 years later, because the western world just would not knowingly allow it. People now object strongly to such inhumane treatment, and in many countries there are laws to prevent it. Of course, the reaction is somewhat irrational, because the animals to be protected (in this case the baboons) probably suffer little more than they would if they were taken in the wild by a wild leopard, without human intervention. But in our present-day society few would argue with the notion that we cannot allow people to cause animal suffering. This sympathy for wild animals is relatively young, and is still increasing.

The concern about cruelty towards and exploitation of animals in western society affects our dealings with all mammals and birds, but perhaps with carnivores more than most. Demonstrations and publicity against vivisection and cruelty often focus on dogs (beagles in nicotine experiments), cats (experiments on brain function) or bears (performing, or caged to extract gall). There are strong public sentiments against the wearing of furs, and in Britain the farming of animals for the fur trade is to be banned within the next few years. The European Union is expected to proscribe trade with any country that still allows the trapping of wild animals for fur with leghold traps. It is possible that the hunting of foxes in Britain is likely to be outlawed soon. The talk is about demands for animal rights.

These changes imply that gradually, animals are to be treated more and more with the respect that we afford other people. The trend is gaining ground in many western countries, and it is difficult to guess at present how far it will go. Ultimately, it may affect what we allow wild animals to do to each other, which is not as far-fetched as it may seem. Many countries ban staged dogfights or cockfights; baiting of bears or badgers with dogs is fortunately now illegal. TV audiences are sensitive about the amount of natural and wild animal violence to be shown on their screens, and films of predator kills have to be carefully edited to make them acceptable. Educationalists are concerned about bloodthirsty beasts in children's stories. From this point onwards it was not a large step to the suggestion from readers of one of George Schaller's popular and graphic accounts of African wild dogs killing wildebeest. Letters to Schaller's editor mooted the idea that this bloody carnage should stop, that the authorities in charge of national parks should feed those repulsive wild dogs with prey killed with a clean shot, instead of allowing the predators to tear zebra and wildebeest apart, slowly and cruelly (Dr G.B. Schaller, pers. comm.).

Today, most of us will still shrug our shoulders over this suggestion as being preposterous, even for the distant future. But with animals more and more confined to small game parks and surrounded by fences, the idea becomes almost feasible. It even becomes attractive to those who see animals as having some sort of rights and wants, as do people, and who wish to transfer human morals to animals (I do not include myself here). The fact that such ideas are even considered today shows just how much our relationship with wild carnivores has evolved over the years. Let me just recap some of our findings from the previous chapters, starting with our joint evolution.

Contemporary carnivores show a variety of about 230 species, usually divided into eight families. But before mankind arrived on the scene there used to be many more species, and these carnivores themselves had replaced earlier orders of mammals of rather similar types. Carnivores dominated the earthly scene, they were at the top of the feeding pyramid before and during the early human evolution, and we evolved whilst suffering their presence. We lived with them, competed with them, defended ourselves against them, and we came to use, and finally to love and admire them.

Most of the carnivores are small, but some are large and dangerous to ourselves, and to our children and livestock. Because all carnivores have so much of their appearance in common we are inclined to treat them all somewhat similarly: deep down, most people consider that carnivores carry at least some danger to us. Our reactions to wild carnivores contain elements of fear, but also of aggression and curiosity, and the balance between these basic ingredients of our behaviour is finely tuned to the ecological significance of the predators to us. We fear the carnivore that threatens us in person, we attack the predator that threatens our children or other dependants, or our livestock, and we are most curious towards the really dangerous ones.

Yet that is just the basic and instinctive part of our behaviour, and it is as well to remember that not all our reactions to these animals are governed by fear and aggression. Some of that curiosity, of the deep attraction we feel and the beauty we see in carnivores is due to our fascination with hunting, and this draws people to watch the animals in the wild or on their TV screens, in combat with prey. We are hunters ourselves, in reality (at least we were not so long ago), or in our dreams and expectations, and the process of prey capture makes compulsive viewing.

As a quite separate and very important development in our relations with carnivores, mankind domesticated and enslaved the dog and the cat. It started as a purely utilitarian arrangement, as cats were kept to control vermin around the house, and dogs were trained for a host of different tasks. Then, through the ages and especially during the last century, this relationship evolved into another. This was a tremendously important process, where the carnivore, the servant, came to be valued in a quite different role, as a companion rather than a worker.

Now, behavioural characteristics that these domesticated animals share with people are being emphasized, especially in western countries. We interact with them, often almost in the same way as we interact with people. Pets are being appreciated in a manner quite different from before, when our dogs and cats still had to work for a living. We are seeing them through different eyes, and the image we perceive necessarily rubs off on the way in which we treat the animals we keep for some utilitarian purpose, and on the way in which we see wild animals, especially the wild brethren of dogs and cats, the other carnivores.

Around us, on the one hand we see a decline in wilderness and in wild populations. Overall, there is a decreasing physical threat from wild carnivores to ourselves and our livestock, and we are less exposed to them. There are fewer occasions when we interact with wild animals when we are the potential prey, and also fewer times when wild carnivores take our livestock. The danger is receding.

On the other hand, we have better opportunities directly (from a vehicle through binoculars) or indirectly (through films and TV) to see exactly what happens during the animal hunt, when we are so close to events that we almost become personally involved. Some of us may have strong urges to protect the prey, or people like me identify closely with the hunter. Moreover, as a separate development, many of us are seeing wild animals more as individuals, we are humanizing them somewhat as an extension of our relationships with pets, and we are becoming more averse to their violence.

Looking further ahead, my grandchildren will almost inevitably see carnivores very differently from the way that I do. They will be attracted perhaps just as much, but for different reasons. There will be fewer wild animals, less wilderness, but much of what there is will be appreciated then as 'nature monuments', fenced in and safe. Some of the more exciting ecosystems with large predators will be preserved, we hope, as we now preserve works of art. The magic of dangerous carnivores will have disappeared, and a more cultivated image will have taken its place.

The effects of carnivores on our society have been far-reaching, in fact much further than appears at first sight. On the miserable side of the balance there are some observed effects that may be horrific, as predators cause mortality and damage through direct attacks and disease, and as they take our livestock and game; and then there are the indirect effects, the time and energy we spend in preventing damage, precautions we take almost without thinking about it. There is the psychological cost, the fear and nightmares caused by the mere presence

Serval

of the animals. But on the positive side there is the admiration that we foster for them because we were hunters, and the thrill we derive from just the sight of these animals. There is the large influence of carnivores on our culture, on our literature, fables and folk tales, heraldry, even witchcraft.

Many of these aspects of our carnivore connection are likely to be affected by the current process of change. One is more aware of wildlife because of films on TV, more aware of the individual animals that make up populations. At least part of this change is generated by our relations with cats and dogs at home. It has to be quite likely, therefore, that our developing relationship with pets will have a profound influence on conservation, in the future even more than it had in the past.

There often is contempt amongst the diehard conservationists for the sentimentality of people towards pets. But I think that such contempt is misplaced, and it could be counterproductive. Respect and love for the cat and dog at home is bound to foster respect and love for the wolf, the bear, the fox and the leopard, and appreciating the needs of pets cannot help but make an impact on the conservation of wild animals. I firmly believe that more people should be encouraged to keep pets, and if properly guided, the interest in pets could lead to benefits for conservation projects. Of course, there is a danger that the concern with individual domestic animals will lead to a thoughtless extrapolation of human values and morals onto animals. That could lead to excesses such as those already mentioned, including the suggested provisioning of wild carnivores with cleanly killed prey, or the culling of 'immoral killers', and other freakish trends.

However, I am confident that such danger can be kept in check if people are made familiar with the information that behavioural and ecological science can provide, and if one can have the chance to see for oneself how animals treat animals, if only on TV. The process of public perception of wild animals needs guidance, and I believe that ecologists and conservationists should provide it. The way in which we view individual animals, and tolerate their behaviour, is fundamental to our relationships with all the fauna. Mankind's interaction with carnivores deserves to be a model. Carnivores, in their prominent position at the top of the food pyramid, may sometimes be a menace to our existence, but as helpers to mankind and as beautiful hunter idols, they stand as symbols for the wildlife that surrounds us.

References

Ardrey, R. (1966). *The Territorial Imperative.* New York: Atheneum.

Attwell, R.I.G. (1958). The African hunting dog: a wildlife management incident. *Oryx*, **4**, 326–8.

Baerends, G.P. (1970). A model of the functional organization of incubation behaviour. *Behaviour, Suppl.*, **17**, 263–312.

Bailey, E.P. (1982). Effects of fox farming on Alaskan islands and the proposed use of red foxes as biological control agents for introduced Arctic foxes. *Bull. Pac. Seabird Group*, **9**, 74–5.

Bailey, T.N. (1993). *The African Leopard.* New York: Columbia University Press.

Baker, R.O. & Timm, R.M. (1998). Management of conflicts between urban coyotes and humans in southern California. In *Proceedings, 18th Vertebrate Pest Conference*, ed. R.O. Baker & A.C. Crabb, pp. 299–312. Davis: University of California.

Balestra, F.A. (1962). The man-eating hyenas of Mlanje. *Afr. Wildl.*, **16**, 25–7.

Ballard, W.B., Ayres, L.A., Krausman, P.R., Reed, D.J. & Fancy, S.G. (1997). Ecology of wolves in relation to a migratory caribou herd in northwest Alaska. *Wildl. Monogr.*, **135**, 5–47.

Balmford, A., Mace, G.M. & Leader-Williams, N. (1996). Designing the ark: setting priorities for captive breeding. *Conserv. Biol.*, **10**, 719–27.

Baltz, M.E. & Ratnaswamy, M.J. (2000). Mascot conservation programs: using college animal mascots to support species conservation efforts. *Wildl. Soc. Bull.*, **28**, 159–63.

Bartmann, R.M., White, G.C. & Carpenter, L.H. (1992). Compensatory mortality in a Colorado mule deer population. *Wildl. Monogr.*, **121**, 1–39.

Beier, P. (1991). Cougar attacks on humans in the United States and Canada. *Wildl. Soc. Bull.*, **19**, 403–12.

Bekoff, M. & Wells, M.C. (1986). Social ecology and behavior of coyotes. *Adv. Study Behav.*, **16**, 251–338.

Berger, J. (1998). Future prey: some consequences of the loss and restoration of large carnivores. In *Behavioral Ecology and Conservation Biology*, ed. T. Caro, pp. 80–103. Oxford: Oxford University Press.

Berger, J., Swenson, J.E. & Persson, I.L. (2001). Recolonizing carnivores and naïve prey: conservation lessons from Pleistocene extinctions. *Science*, **291**, 1036–9.

Berke, O. & von Keyserlingk, M. (2001). Increase of the prevalence of *Echinococcus multilocularis* infection in red foxes in Lower Saxony. *Deut. Tierarztl. Wochenschr.*, **108**, 201–5.

Bininda-Emonds, O.R., Gittleman, J.L. & Purvis, A. (1999). Building large trees by combining phylogenetic information: a complete phylogeny of the extant Carnivora (Mammalia). *Biol. Rev.*, **74**, 143–75.

Bodner, M. (1998). Damage to fish ponds as a result of otter (*Lutra lutra*) predation. *BOKU Rep. on Wildl. Res. & Game Manage. (Vienna)*, **14**, 106–17.

Boschetti, A. & Kasznica, J. (1995). Visceral larva migrans induced eosinophilic cardiac pseudotumor: a cause of sudden-death in a child. *J. Forensic Sci.*, **40**, 1097–9.

Bourliere, F. (1963). Specific feeding habits of African carnivores. *Afr. Wildl.*, **17**, 21–7.

Bourne, W.R.P. (1965). The missing petrels. *Bull. Brit. Ornithol. Club*, **85**, 97–105.

Bowen, W.D. (1982). Home range and spatial organisation of coyotes in Jasper National Park, Alberta. *J. Wildl. Manage.*, **46**, 201–16.

Brace, M. (1998). Ill Met by Moonlight. *The Independent*, 20 June 1998.

Brain, C.K. (1981). *The Hunters or the Hunted?* Chicago: University of Chicago Press.

Broekhuizen, S. (1983). Habitat use of beech marten (*Martes foina*) in relation to landscape elements in a Dutch agricultural area. In *Proceedings of the XVI Congress of the International Union of Game Biologists*, 614–24.

Brotherton, P.N.M., Clutton-Brock, T.H., O'Riain, M.J., Gaynor, D., Sharpe, L., Kansky, R. & McIlrath, G.M. (2001). Offspring food allocation by parents and helpers in a cooperative mammal. *Behav. Ecol.*, **12**, 590–9.

Bryant, B.K. (1990). The richness of the child-pet relationship: a consideration of both benefits and costs of pets to children. *Anthrozoos*, **3**, 253–61.

Burrows, R. (1995). Demographic changes and social consequences in wild dogs 1964–1992. In *Serengeti II: Research, Management and Conservation of an Ecosystem*, ed. A.R.E. Sinclair & P. Arcese, pp. 400–20. Chicago: University of Chicago Press.

Burt, W.H. & Grossenheider, R.P. (1959). *A Field Guide to the Mammals*. Boston: Houghton Mifflin.

Campbell, K. & Hofer, H. (1995). People and wildlife: spatial dynamics and zones of interaction. In *Serengeti II: Dynamics, Management and Conservation of an Ecosystem*, ed. A.R.E. Sinclair & P. Arcese, pp. 534–70. Chicago: University of Chicago Press.

Carbone, C., Mace, G.M., Roberts, S.G. & Macdonald, D.W. (1999). Energetic constraints on the diet of terrestrial carnivores. *Nature*, **402**, 286–88.

Carbyn, L.N. (1989). Coyote attacks on children in western North America. *Wildl. Soc. Bull.*, **17**, 444–6.

Caro, T.M. (1994). *Cheetahs of the Serengeti Plains*. Chicago: University of Chicago Press.

Carss, D.N. & Elston, D.A. (1996). Errors associated with otter *Lutra lutra* faecal analysis. *J. Zool.*, **238**, 319–32.

Carter, S.K. & Rosas, F.C.W. (1997). Biology and conservation of the giant otter *Pteronura brasiliensis*. *Mammal Rev.*, **27**, 1–26.

Clark, K. (1977). *Animals and Men: Their Relationship as Reflected in Western Art from Prehistory to the Present Day*. London: Thames & Hudson.

Clark, T.W. (1989). Conservation biology of the black-footed ferret, *Mustela nigripes*. *Wildl. Preserv. Trust Spec. Sci. Rep.*, **3**, 1–175.

Cleaveland, S. & Dye, C. (1995). Maintenance of a microparasite infecting several host species: rabies in the Serengeti. *Parasitology*, **111**, 33–47.

Clevenger, A.P., Campos, M.A. & Hartasanchez, A. (1994). Brown bear *Ursus arctos* predation on livestock in the Cantabrian mountains, Spain. *Acta Theriol.*, **39**, 267–78.

Clutton-Brock, J. (1995). Origins of the dog: domestication and early history. In *The Domestic Dog*, ed. J. Serpell, pp. 7–20. Cambridge: Cambridge University Press.

Clutton-Brock, T.H., Brotherton, P.N.M., Russell, A.F., O'Riain, M.J., Gaynor, D., Kansky, R., Griffin, A., Manser, M., Sharpe, L., McIlrath, G.M., Small, T., Moss, A. & Monfort, S. (2001). Cooperation, control and concession in meerkat groups. *Science*, **291**, 478–81.

Clutton-Brock, T.H. & Harvey, P.H. (1984). Comparative aproaches to investigating adaptation. In *Behavioral Ecology: An Evolutionary Approach*, ed. J.R. Krebs & N.B. Davies, pp. 7–29. Sunderland, Mass.: Sinauer.

Clutton-Block, T.H., Orian, M.J., Brotherton, P.N.M., Gaynor, D., Kansky, R., Griffin, A.S. & Manser, M. (1999). Selfish sentinels in cooperative mammals. *Science*, **284**, 1640–4.

Constable, A. (1891). *Travels to the Mogul Empire, A.D. 1656–68*. London: Constable.

Corbett, J. (1991). *The Jim Corbett Omnibus*. New Delhi: Oxford University Press.

Corbett, L. (1995). *The Dingo in Australia and Asia*. Sydney: UNSW Press.

Cozza, K., Fico, R., Battistini, M. & Rogers, E. (1996). The damage-conservation interface illustrated by predation on domestic livestock in central Italy. *Biol. Conserv.*, **78**, 329–36.

Craik, J.C.A. (1998). Recent mink-related declines of gulls and terns in west Scotland and the beneficial effects of mink control. *Argyll Bird Rep.*, **14**, 98–110.

Creel, S. & Creel, N.M. (1995). Communal hunting and pack size in African wild dogs, *Lycaon pictus*. *Anim. Behav.*, **50**, 1325–39.

Creel, S.R. & Waser, P.M. (1994). Inclusive fitness and reproductive strategies in dwarf mongooses. *Behav. Ecol.*, **5**, 339–48.

Crook, K.R. & Soule, M.E. (1999). Mesopredator release and avifaunal extinctions in a fragmented system. *Nature*, **400**, 563–6.

Darwin, C. (1859). *On the Origin of Species by Means of Natural Selection, or the Preservation of Favourite Races in the Struggle for Life*. London: John Murray.

Delpietro, H., Konolsaisen, F., Marchevsky, N. & Russo, G. (1994). Domestic cat predation on vampire bats (*Desmodus rotundus*) while foraging on goats, pigs, cows and human beings. *Appl. Anim. Behav. Sci.*, **39**, 141–50.

De Luca, D.W. & Ginsberg, J.R. (2001). Dominance, reproduction and survival in banded mongooses: towards an egalitarian social system? *Anim. Behav.*, **61**, 17–30.

Derenne, P. & Mougin, J.L. (1976). Données écologiques sur les mammiferes introduits de l'Ile aux Cochons, Archipel Crozet (46° 06′ S, 50° 14′ E). *Mammalia*, **40**, 21–52.

Dickman, C.R. (1996). Impact of exotic generalist predators on the native fauna of Australia. *Wildl. Biol.*, **2**, 185–95.

Doncaster, C.P. (1992). Testing the role of intraguild predation in regulating hedgehog populations. *Proc. Roy. Soc. B*, **249**, 113–17.

Dragoo, J.W. & Honeycutt, R.L. (1997). Systematics of mustelid carnivores. *J. Mammal.*, **78**, 426–43.

Dunnet, G.M., Jones, D.M. & McInerney, J.P. (1986). *Badgers and Bovine Tuberculosis*. London: HMSO.

Dunstone, N. (1993). *The Mink*. London: T. & A.D. Poyser.

East, M.L. & Hofer, H. (1991). Loud-calling in a female-dominated mammalian Society: I, II. *Anim. Behav.*, **42**, 637–49, 651–69.

Eckert, J., Conraths, F.J. & Tackmann, K. (2000). Echinococcosis: an emerging or re-emerging zoonosis? *Int. J. Parasitol.*, **30**, 1283–94.

Elton, C. & Nicholson, M. (1942). The ten-year cycle in numbers of the lynx in Canada. *J. Anim. Ecol.*, **11**, 215–44.

Eltringham, S.K. (1984). *Wildlife Resources and Economic Development*. Chichester: Wiley & Sons.

Encyclopaedia Brittanica (2002). Bestiary. http:www.britannica.com

Errington, P.L. (1946). Predation and vertebrate populations. *Q. Rev. Biol.*, **21**, 144–77, 221–45.

Estes, J.A., Tinker, M.T., Williams, T.M. & Doak, D.F. (1998). Killer whale predation on sea otters linking oceanic and nearshore ecosystems. *Science*, **282**, 473–6.

Estes, R.D. (1991). *The Behaviour Guide to African Mammals*. Berkeley: University of California Press.

Fico, R., Morosetti, G. & Giovanni, A. (1993). The impact of predators on livestock in the Abruzzo region of Italy. *Rev. Sci. Tech. Off. Int. Epiz.*, **12**, 39–50.

Fitzgibbon, C.D. (1994). The costs and benefits of predator inspection behavior in Thomson's gazelles. *Behav. Ecol. Sociobiol.*, **34**, 139–48.

Fleck, S. & Herrero, S. (1989). Polar bear conflicts with humans. In *Bear–People Conflicts*: Proceedings of a Symposium on Management Strategies, ed. M. Bromley. Yellowknife, Yukon: Northwest Territories Department of Renewable Resources.

Flynn, J. J. (1996). Carnivoran phylogeny and rates of evolution: morphological, taxic and molecular. In *Carnivore Behavior, Ecology and Evolution II*, ed. J.L. Gittleman, pp. 542–81. Ithaca, N.Y.: Cornell University Press.

Forsell, D.J. (1982). Recolonization of Baker Island by seabirds. *Bull. Pac. Seabird Group*, **9**, 75–6.

Foster-Turley, P. (1998). Fishing with otters – a fading tradition. *Oryx*, **32**, 2.

Frame, L.H., Malcolm, J.R., Frame, G.W. & van Lawick, H. (1979). Social organization of African wild dogs (*Lycaon pictus*) on the Serengeti plains. *Z. Tierpsychol.*, **50**, 225–49.

Frank, L.G. (1986). Social organization of the spotted hyena (*Crocuta crocuta*). II. Dominance and reproduction. *Anim. Behav.*, **35**, 1510–27.

Friar, S. (1996). *Heraldry for the Local Historian and Genealogist*. London: Sutton.

Fritts, S.H., Paul, W.J., Mech, L.D. & Scott, D.P. (1992). Trends and management of wolf-livestock conflicts in Minnesota. *Report* 181 Washington, D.C.: US Department of the Interior Fish & Wildlife Service.

Fritz, C.L., Farver, T.B., Kass, P.H. & Hart, L.A. (1995). Association with companion animals and the expression of noncognitive symptoms in Alzheimer's patients. *J. Nerv. Ment. Dis.*, **183**, 459–63.

Fritzell E.K. (2001). Raccoons. In *The New Encyclopedia of Mammals*, ed. D. Macdonald, pp. 88–9. Oxford: Oxford University Press.

Fuller, T.K., Kat, P.W., Maddock, A.H., Ginsberg, J.R., Burrows, R., McNutt, J.W. & Mills, M.G.L. (1992a). Population dynamics of African wild dogs. In *Wildlife 2001: Populations*, ed. D.R. McCullough & H. Barrett, pp. 1125–39. London: Elsevier.

Fuller, T.K., Mills, M.G.L., Borner, M., Laurenson, K. & Kat, P.W. (1992b). Long distance dispersal by African wild dogs in East and South Africa. *J. Afr. Zool.*, **106**, 535–7.

Funderbunk, S. (1986). International trade in United States and Canadian bobcats, 1977–81. In *Cats of the World: Biology, Conservation and Management*, ed. S.D. Miller & D.D. Everell, pp. 489–501. Washington D.C.: National Wildlife Federation.

Gasaway, W.C., Boertje, R.D., Grangaard, D.V., Kelleyhouse, D.G., Stephenson, R.O. & Larsen, D.G. (1992). The role of predation in limiting moose at low densities in Alaska and Yukon and implications for conservation. *Wildl. Monogr.*, **120**, 1–59.

Geraerdts, G. (1981). Is angst voor wolven terecht? *De Maasgouw*, **100**, 193–204.

Gesner, K. (1551–87). *Historiae Animalium*. Zurich: Christoph Froschauer.

Gittleman, J.L. (1985). Carnivore body size: ecological and taxonomic correlates. *Oecologia*, **67**, 540–54.

Gittleman, J.L. (1989). Carnivore group living: comparative trends. In *Carnivore Behavior, Ecology and Evolution*, ed. J.L. Gittleman, pp. 183–208. Ithaca, N.Y.: Cornell University Press.

Gompper, M.E. (1996). Sociality and asociality in white-nosed coatis (*Nasua narica*): foraging coasts and benefits. *Behav. Ecol.*, **7**, 254–63.

Gompper, M.E. (1997). Population ecology of the white-nosed coati (*Nasua narica*) on Barro Colorado Island, Panama. *J. Zool.*, **241**, 441–55.

Gompper, M.E., Gittleman, J.L. & Wayne, R.K. (1997). Genetic relatedness, coalitions and social behaviour of white-nosed coatis, *Nasua narica*. *Anim. Behav.*, **53**, 781–97.

Gompper, M.E., Gittleman, J.L. & Wayne, R.K. (1998). Dispersal, philopatry, and genetic relatedness in a social carnivore: comparing males and females. *Molec. Ecol.*, **7**, 157–63.

Goodall, J. (1971). *In the Shadow of Man*. London: Collins.

Gorman, M., Mills, M.G., Raath, J. & Speakman, J.R. (1998). High hunting costs make African wild dogs vulnerable to kleptoparasitism by hyaenas. *Nature*, **397**, 479–81.

Gosling, L.M. (1982). A reassessment of the function of scent marking in territories. *Z. Tierpsychol.*, **60**, 89–118.

Green, J. & Woodruff, R. (1984). Livestock-guarding dogs for predator control: costs, benefits, and practicality. *Wildl. Soc. Bull.*, **12**, 44–50.

Griffiths, H.I. (1993). The Eurasian badger (*Meles meles*) (L. 1758) as a commodity species. *J. Zool.*, **230**, 340–42.

Grimm, J. & Grimm, W. (1957). *Die Märchen der Brüder Grimm*. München: Goldmann Verlag.

Guggisberg, C.A.W. (1962). *Simba, the Life of the Lion*. London: Bailey Bros. & Swinfen.

Harris, S. (1986). *Urban Foxes*. London: Whittet Books.

Hart, L.A. (1995). Dogs as human companions: a review of the relationship. In *The Domestic Dog*, ed. J. Serpell, pp. 161–78. Cambridge: Cambridge University Press.

Heinen, J.T. & Leisure, B. (1993). A new look at the Himalayan fur trade. *Oryx*, **27**, 231–8.

Heinsohn, R. & Packer, C. (1995). Complex cooperative strategies in group-territorial African lions. *Science*, **269**, 1260–2.

Hendrichs, H. (1975). The status of the tiger *Panthera tigris* (Linné 1758) in the Sundarbans mangrove forest (Bay of Bengal). *Säugetierk. Mitt.*, **23**, 161–99.

Henry, J.D. (1977). The use of urine marking in the scavenging behaviour of the red fox (*Vulpes vulpes*). *Behaviour*, **61**, 82–105.

Herrero, S. (1985). *Bear Attacks: Their Causes and Avoidance*. Piscataway, N.J.: Winchester Press.

Hewson, R. (1981). The red fox as a scavenger and predator of sheep in west Scotland. PhD thesis, Aberdeen: University of Aberdeen.

Hilton-Taylor, C. (Compiler) (2000). *2000 IUCN Red List of Threatened Species*. Gland: IUCN.

Hofer, H. (1998). Spotted hyaena *Crocuta crocuta* (Erxleben, 1777). In *Hyaenas. Status Survey and Conservation Action Plan*, ed. M.G.L. Mills & H. Hofer, pp. 29–38. Gland: IUCN.

Holekamp, K.E., Ogutu, J.O., Dublin, H.T., Frank, L.G. & Smale, L. (1993). Fission of a spotted hyaena clan: consequences of prolonged female absenteeism and causes of female emigration. *Ethology*, **93**, 285–99.

Holekamp, K.E. & Smale, L. (1990). Provisioning and food-sharing by lactating spotted hyenas, *Crocuta crocuta* (Mammalia, Hyaenidae). *Ethology*, **86**, 191–202.

Horn, E. (1964). *Wild in der Küche*. München: Mayer Verlag.

Howard, H.E. (1920). *Territory in Bird Life*. London: John Murray.

Hudson, P.J. (1992). *Grouse in Space and Time*. Fordingbridge, UK: Game Conservancy.

Hunt, R.M. (1996). Biogeography of the order Carnivora. In *Carnivore Behavior, Ecology and Evolution II*, ed. J.L. Gittleman, pp. 485–541. Ithaca: Cornell University Press.

Huntingford, F.A. (1976). The relationship between inter- and intra-specific aggression. *Anim. Behav.*, **24**, 485–97.

IUCN (1996). Puma (*Puma concolor*). http://lynx.uio.no/catfolk/puma-06.htm.

Jenkins, D., Watson, A. & Miller, G.R. (1964). Predation and red grouse populations. *J. Appl. Ecol.*, **1**, 183–95.

Jenkins, J.L. (1986). Physiological effects of petting a companion animal. *Psychol. Rep.*, **58**, 21–2.

Johnsingh, A.J.T. (1982). Reproductive and social behaviour of the dhole (*Cuon alpinus*). *J. Zool.*, **198**, 443–63.

Johnston, H.A.S. (1966). *A Selection of Hausa Stories*. Oxford: Clarendon Press.

Jones, J.L., Kruszon-Moran, D., Wilson, M., McQuillan, G., Navin, T. & McAuley, B. (2001). *Toxoplasma gondii* infection in the United States: seroprevalence and risk factors. *Am. J. Epidemiol.*, **154**, 357–65.

Jones, M.E. (1995). Guild structure of the large marsupial carnivores in Tasmania. PhD Thesis. Hobart: University of Tasmania.

Jorga, W. (1998). *Der Lausitzer Wassermann lebt*. Berlin: Cottbuser Bücher.

Kaplan, C. (1977). *Rabies, the Facts*. Oxford: Oxford University Press.

Karanth, K.U. & Sunquist, M.E. (2000). Behavioural correlates of predation by tiger (*Panthera tigris*), leopard (*Panthera pardus*) and dhole (*Cuan alpinus*) in Nagarahole, India. *J. Zool.*, **250**, 255–65.

Kaufman, J.H. (1962). Ecology and social behavior of the coati, *Nasua narica*, on Barro Colorado Island, Panama. *Univ. Calif. Publ. Zool.*, **60**, 95–222.

Kays, R.W. & Gittleman, J.L. (2001). The social organization of the kinkajou *Potamus flavus* (Procyonidae). *J. Zool.*, **253**, 491–504.

Keane, B., Creel, S.R. & Waser, P.M. (1996). No evidence of inbreeding avoidance or inbreeding depression in a social carnivore. *Behav. Ecol.*, **7**, 480–9.

Kingdon, J. (1997). *The Kingdon Field Guide to African Mammals*. London: Academic Press.

Kipling, R. (1994). *The Jungle Book*. London: Puffin.

Kleiman, D.G. & Eisenberg, J.F. (1973). Comparisons of canid and felid social systems from an evolutionary perspective. *Anim. Behav.*, **21**, 637–59.

Knarrum, V., Sorensen, O.J., Eggen, T., Kvam, T. & Opseth, O. (1996). Brown bear prey selection on domestic sheep in Norway. In *2nd International Symposium on the Coexistence of Large Carnivores with Man, Saitama, Japan*, Poster. Tokyo: Ecosystem Conservation Society.

Koganezawa, M., Maruyama, N., Takahashi, M., Chinen, S. & Angeli, C. (1996). Japanese people's attitudes toward wolves and their reintroduction into

Japan. In *2nd International Symposium on the Coexistence of Large Carnivores with Man*, Saitama, Japan, abstract, p. 124. Tokyo: Ecosystem Conservation Society.

Kohl-Larsen, L. (1956). *Das Zauberhorn, Märchen und Tiergeschichten der Tindiga*. Kassel: Röth Verlag.

Kolb, H.H. & Hewson, R. (1980). A study of fox populations in Scotland from 1971 to 1976. *J. Appl. Ecol.*, **17**, 7–19.

Korpimaki, E. & Norrdahl, K. (1998). Experimental reduction of predators reverses the crash phase of small rodent cycles. *Ecology*, **79**, 2448–55.

Kothari, A., Pande, P., Singh, S. & Variava, D. (1989). *Management of National Parks and Sanctuaries in India: A Status Report*. Delhi: Indian Institute of Public Administration.

Kruuk, H. (1964). Predators and anti-predator behaviour of the black-headed gull, *Larus ridibundus*. *Behaviour*, **Suppl. 11**, 1–129.

Kruuk, H. (1965). Masai art. *Animals*, **6**, 494–9.

Kruuk, H. (1967). Competition for food between vultures in East Africa. *Ardea*, **55**, 171–93.

Kruuk, H. (1972a). *The Spotted Hyena*. Chicago: University of Chicago Press.

Kruuk, H. (1972b). Surplus killing in carnivores. *J. Zool.*, **166**, 233–44.

Kruuk, H. (1975). *Hyaena*. Oxford: Oxford University Press.

Kruuk, H. (1976a). The biological function of gulls' attraction towards predators. *Anim. Behav.*, **24**, 146–53.

Kruuk, H. (1976b). Feeding and social behaviour of the striped hyaena (*Hyaena vulgaris* Desmarest). *E. Afr. Wildl. J.*, **14**, 91–111.

Kruuk, H. (1978a). Spatial organisation and territorial behaviour of the European badger, *Meles meles* L. *J. Zool.*, **184**, 1–15.

Kruuk, H. (1978b). Foraging and spatial organization of the European badger, *Meles meles* L. *Behav. Ecol. Sociobiol.*, **4**, 75–89.

Kruuk, H. (1980). The effects of large carnivores on livestock and animal husbandry in Marsabit District, Kenya. *IPAL Report* E-4. Nairobi: United Nations Environmental Programme.

Kruuk, H. (1986). Interactions between Felidae and their prey species: a review. In *Cats of the World: Biology, Conservation and Management*, ed. S.D. Miller & D.D. Everell, pp. 353–73. Washington D.C.: National Wildlife Federation.

Kruuk, H. (1989). *The Social Badger*. Oxford: Oxford University Press.

Kruuk, H. (1995). *Wild Otters: Predation and Populations*. Oxford: Oxford University Press.

Kruuk, H. & de Kock, L. (1981). Food and habitat of badgers, *Meles meles*, on Monte Baldo, N. Italy. *Z. Säugetierkd.*, **46**, 295–301.

Kruuk, H., Kanchanasaka, B., O'Sullivan, S. & Wanghongsa, S. (1994). Niche separation in three sympatric otters *Lutra perspicillata*, *L. lutra* and *Aonyx cinerea*, in Huay Kha Khaeng, Thailand. *Biol. Conserv.*, **69**, 115–20.

Kruuk, H. & Moorhouse, A. (1991). The spatial organization of otters (*Lutra lutra* L.) in Sdhetland. *J. Zool.*, **224**, 41–57.

Kruuk, H. & Parish, T. (1982). Factors affecting population density, groupsize and territory size in the European badger, *Meles meles*. *J. Zool.*, **196**, 31–9.

Kruuk, H. & Snell, H. (1981). Prey selection by feral dogs from a population of marine iguanas. *J. Appl. Ecol.*, **18**, 197–204.

Kugelschafter, K., Rothe, H. & Wiesmaier, K. (1993). Zur lokalen Ausbreitung der sogenannten Automarder-Schäden (*Martes foina* Erxleben, 1777). *Z. Säugetierkd.*, **58 (Sonderheft)**, 40–1.

Kurten, B. (1968). *Pleistocene Mammals of Europe*. London: Weidenfeld & Nicholson.

Kvitek, R.G. & Oliver, J.S. (1988). Sea otter foraging habits and effects on prey populations and communities in soft-bottom environments. In *The Community*

Ecology of Sea Otters, ed. G.R. VanBlaricom & J.A. Estes, pp. 22–47. Berlin: Springer.

Lack, D. (1954). *The Natural Regulation of Animal Numbers*. Oxford: Oxford University Press.

La Fontaine, J. (1997). *Fables*. Paris: Bookking International, Classiques française.

Landa, A., Gudvangen, K., Swenson, J.E. & Røskaft, E. (1999). Factors associated with wolverine *Gulo gulo* predation on domestic sheep. *J. Appl. Ecol.*, **36**, 963–73.

Landa, A., Strand, O., Swenson, J.E. & Skogland, T. (1997). Wolverines and their prey in southern Norway. *Can. J. Zool.*, **75**, 1292–9.

Landa, A. & Tommerås, B.Å. (1996). Do volatile repellents reduce wolverine predation on sheep? *Wildl. Biol.*, **2**, 119–26.

Leakey, M.G. (1976). Carnivora of the East Rudof succession. In *Earliest Man and Environments in the Lake Rudolf Basin*, ed. Y. Coppens, F.C. Howell, G.L. Isaac & R.E.F. Leakey, pp. 302–313. Chicago: University of Chicago Press.

Leakey, M.G. (1983). *Africa's Vanishing Art, the Rock Paintings of Tanzania*. New York: Doubleday.

Leakey, R. (1994). *The Origin of Humankind*. London: Weidenfeld & Nicholson.

Leakey, R. & Lewin, R. (1979). *People of the Lake*. London: Collins.

Leckie, F.M., Thirgood, S.J., May, R. & Redpath, S.M. (1998). Variation in the diet of red foxes on moorland in relation to prey abundance. *Ecography*, **21**, 599–604.

Lenaghan, R.T. (Editor) (1967). *Caxton's Aesop*. Cambridge, Mass.: Harvard University Press.

Lewis, M. (Editor) (1997). *The Journals of Lewis and Clark*. Boston: Mariner Books.

Lhote, H. (1958). *À la Découverte des Fresques du Tassili*. Paris: Editions Arthaud.

Lindström, E.R., Andren, H., Angelstam, P., Cederlund, G., Hörnfeldt, B., Jäderberg, L., Lemnell, P., Martinsson, B., Sköld, K. & Swenson, J.E. (1994). Disease reveals the predator: sarcoptic mange, red fox predation, and prey populations. *Ecology*, **75**, 1042–9.

Lingle, S. (2000). Seasonal variation in coyote feeding behaviour and mortality of white-tailed deer and mule deer. *Can. J. Zool.*, **78**, 85–99.

Longhurst, W., Leopold, S. & Dasman, R. (1952). A survey of Californian deer herds, their ranges and management problems. *Bulletin No. 6*. Sacramento: California Division of Natural Resources, Department of Fish & Game.

Loveridge, A.J. & Macdonald, D.W. (2001). Seasonality in spatial organization and dispersal of sympatric jackals (*Canis mesomelas* and *C. adustus*): implications for rabies management. *J. Zool.*, **253**, 101–11.

Macdonald, D.W. (1976). Food caching by red foxes and some other carnivores. *Z. Tierpsychol.*, **42**, 170–85.

Macdonald, D.W. (1980). *Rabies and Wildlife*. Oxford: Oxford University Press.

Macdonald, D.W. (1983). The ecology of carnivore social behaviour. *Nature*, **301**, 379–84.

Macdonald, D.W. (1987). *Running with the Fox*. London: Unwin Hyman.

Macdonald, D.W. (1992). *The Velvet Claw*. London: BBC Books.

Macdonald, D.W. (1995). *European Mammals: Evolution and Behaviour*. London: HarperCollins.

Macdonald, D.W. (2001). *The Encyclopedia of Mammals*. Oxford: Oxford University Press.

Macdonald, D.W. & Barrett, P. (1993). *Mammals of Britain and Europe*. London: HarperCollins.

Macdonald, D.W. & Boitani, L. (1979). Carnivore management: a plea for an ecological ethic. In *Animal Rights*, ed. W. Patterson & R. Ryder, pp. 165–77. London: Centaur Press.

Macdonald, D.W. & Carr, G. (1981). Foxes beware: you are back in fashion. *New Sci.*, **1981**, 9–11.

Mace, G.M. (1994). Classifying threatened species: means and ends. *Philos. Trans. R. Soc. Lond. B Biol. Sci.*, **344**, 91–7.

Maran, T. & Henttonen, H. (1995). Why is the European mink (*Mustela lutreola*) disappearing? A review of the process and hypotheses. *Acta Zool. Fenn.*, **32**, 47–54.

Martin, L.D. (1989). Fossil history of the terrestrial carnivora. In *Carnivore Behavior, Ecology and Evolution*, ed. J.L. Gittleman, pp. 536–68. Ithaca, N.Y.: Cornell University Press.

Martin, R., Rodriguez, A. & Delibes, M. (1995). Local feeding specialization by badgers (*Meles meles*) in a Mediterranean environment. *Oecologia*, 101, 45–50.

Martinek, K., Kolarova, L. & Cerveny, J. (2001a). *Echinococcus multilocularis* in carnivores from the Klatov district of the Czech Republic. *J. Helminthol.*, **75**, 61–6.

Martinek, K., Kolarova, L., Hapl, E., Literak, I. & Uhrin, M. (2001b). *Echinococcus multilocularis* in European wolves (*Canis lupus*). *Parasitol. Res.*, **87**, 838–9.

Maxwell, T.S. (1997). *The Gods of Asia*. Delhi: Oxford University Press.

Mduma, S.A.R., Sinclair, A.R.E. & Hilborn, R. (1999). Food regulates the Serengeti wildebeest: a 40-year record. *J. Anim. Ecol.*, **68**, 1101–22.

Mech, L.D. (1970). *The Wolf. The Ecology and Behaviour of an Endangered Species.* New York: Natural History Press.

Mech, L.D. (1999). Alpha status, dominance and division of labor in wolf packs. *Can. J. Zool.*, **77**, 1196–203.

Mech, L.D. (2000). Wolf restoration to the Adirondacks: the advantages and disadvantages of public participation in the decision. In *Wolves and Human Communities: Biology, Politics and Ethics*, ed. V.A. Sharpe, B. Norton & S. Donelley, pp. 13–22. Washington D.C.: Island Press.

Mech, L.D., Harper, E.K., Meier, T.J. & Paul, W.J. (2000). Assessing factors that may predispose Minnesota farms to wolf depredations on cattle. *Wildl. Soc. Bull.*, **28**, 623–9.

Mills, G. & Hofer, H. (Editors) (1998). *Hyaenas. Status Survey and Conservation Action Plan*. Gland: IUCN, IUCN/SSC Hyaena Specialist Group.

Mills, M.G.L. (1990). *Kalahari Hyaenas*. London: Unwin Hyman.

Mills, M.G.L. & Shenk, T.M. (1992). Predator-prey relationships: the impact of lion predation on wildebeest and zebra populations. *J. Anim. Ecol.*, **61**, 693–702.

Milo, R.G. (1998). Evidence for hominid predation at Klasies River Mount, South Africa, and its implications for the behaviour of early modern humans. *J. Archaeol. Sci.*, **25**, 99–133.

Moehlman, P.D. (1979). Jackal helpers and pup survival. *Nature*, **277**, 382–3.

Moehlman, P.D. (1986). Ecology of cooperation in canids. In *Ecological Aspects of Social Evolution*, ed. D.I Rubenstein & R.W. Wrangham, pp. 64–86. Princeton, N.J.: Princeton University Press.

Moors, P.J. & Atkinson, I.A.E. (1984). Predation on seabirds by introduced animals, and factors affecting its severity. In *Conservation of Island Birds*, ed. P.J. Moors. *ICBP Tech. Publ.*, **3**, 667–90. London: International Council for Bird Preservation.

Morenz, S. (1992). *Egyptian Religion*. Ithaca, N.Y.: Cornell University Press.

Morris, D. (1967). *The Naked Ape*. London: Cape.

Morris, R. & Morris, D. (1965). *Men and Snakes*. London: Hutchinson.

Neal, E. & Cheeseman, C. (1996). *Badgers*. London: T. & A.D. Poyser.

Newmark, W.D., Manyanza, D.N., Gamassa, D.G. & Sariko, H.I. (1994). The conflict between wildlife and local people living adjacent to protected areas in Tanzania. *Conserv. Biol.*, **8**, 249–55.

Newsome, A.E. & Corbett, L.K. (1977). The effects of native, feral and domestic animals on the productivity of the Australian rangelands. In *The impact of herbivores on arid and semi-arid rangelands. Proceedings, 2nd United States/Australia Rangeland Panel, Perth,* 1972, pp. 331–56. Perth: Australian Rangeland Society.

Newsome, A.E., Parer, I. & Catling, P.C. (1989). Prolonged prey suppression by carnivores – predator removal experiments. *Oecologia*, **78**, 458–67.

Nilsson, G. (1980). *Facts about Furs.* Washington, D.C.: Animal Welfare Institute.

Norton-Griffiths, M. (1996). Property rights and the marginal wildebeest: an economic analysis of wildlife conservation options in Kenya. *Biodivers. Conserv.*, **5**, 1557–77.

Nowell, K. & Jackson, P. (1998). *Wild Cats, Status Survey and Action Plan.* Gland: IUCN.

Oli, M.K., Taylor, I.R. & Rogers, M.E. (1994). Snow leopard (*Panthera uncia*) predation of livestock: an assessment of local perceptions in the Annapurna Conservation Area, Nepal. *Biol. Conserv.*, **68**, 63–8.

Packer, C., Gilbert, D.A., Pusey, A.E. & O'Brien, S.J. (1991). A molecular genetic analysis of kinship and cooperation in African lions. *Nature*, **351**, 562–5.

Packer, C. & Pusey, A.E. (1997). Divided we fall: cooperation among African lions. *Sci. Am.*, **276**, 52–9.

Paddle, R. (2000). *The Last Tasmanian Tiger.* Cambridge: Cambridge University Press.

Palomares, F., Ferreras, P., Travaini, A. & Delibes, M. (1998). Coexistence between Iberian lynx and Egyptian mongooses: estimating interaction strength by structural equation modelling and testing by an observational study. *J. Anim. Ecol.*, **67**, 967–8.

Paolo, C. & Boitani, L. (1996). Wolf-livestock conflicts in Tuscany, Italy. In *2nd International Symposium on the Coexistence of Large Carnivores with Man. Saitama, Japan,* abstract, p. 142. Tokyo: Ecosystem Conservation Society.

Patterson, J.H. (1907). *The Man-eaters of Tsavo.* London: Macmillan & Co.

Pemberton, D. (1990). Social organisation and behaviour of the Tasmanian devil *Sarcophilus harrisii.* PhD. Thesis. Hobart: University of Tasmania.

Peterson, R.O. (1999). Wolf-moose interaction on Isle Royale: the end of natural regulation? *Ecol. Appl.*, **9**, 10–16.

Pine, L.G. (1956). *Burke's Genealogical and Heraldic History of the Peerage, Baronetage and Knightage.* London: Burke's Peerage Ltd.

Planetsave.com (2001). Nepal's tigers grow by 50. http://www.planetsave.com/ ViewStory.asp?ID=932

Polis, G.A. & Holt, R.D. (1992). Intraguild predation: the dynamics of complex trophic interactions. *Trends Ecol. Evol.*, **7**, 151–4.

Pons, J.M., Volobouev, V., Ducroz, J.F., Tillier, A. & Reudet, D. (1999). Is the Guadeloupean raccoon (*Procyon minor*) really an endemic species? New insights from molecular and chromosomal analyses. *J. Zool. Syst. Evol. Res.*, **37**, 101–8.

Poortvliet, R. (1994). *Journey to the Ice Age.* New York: Harry N. Abrams.

Post, E., Peterson, R.O., Stenseth, N.C. & McLaren, B.E. (1999). Ecosystem consequences of wolf behavioural response to climate. *Nature*, **401**, 905–7.

Promberger, C. (1996). Carpathian wolf project – field research in Romania. In *2nd International Symposium on the Coexistence of Large Carnivores with Man, Saitama, Japan,* abstract, p. 188. Tokyo: Ecosystem Conservation Society.

Purvis, A., Agapow, P.M., Gittleman, J.L. & Mace, G.M. (2000a). Nonrandom extinction and the loss of evolutionary history. *Science*, **288**, 328–30.

Purvis, A., Gittleman, J.L., Cowlishaw, G. & Mace, G.M. (2000b). Predicting extinction risk in declining species. *Proc. R. Soc. Lond. B, Biol. Sci.*, **267**, 1947–52.

Pye-Smith, C. (1997). *Fox-Hunting: Beyond the Propaganda.* Oakham, Rutland, UK: Wildlife Network.

Quiller-Couch, A. (ed.) (1949). The Oxford Book of English Verse 1250–1918. Clarendon, Oxford.

Ralls, K., Demaster, D.P. & Estes, J.A. (1996). Developing a criterion for delisting the southern sea otter under the US Endangered Species Act. *Conserv. Biol.*, **10**, 1528–37.

Rasa, A. (1984). *Mongoose Watch*. London: John Murray.

Reading, R.P. & Clark, T.W. (1996). Carnivore reintroductions: an interdisciplinary examination. In *Carnivore Behavior, Ecology and Evolution II*, ed. J.L. Gittleman, pp. 296–336. Ithaca, N.Y.: Cornell University Press.

Reed, K.E. (1997). Early hominid evolution and ecological change through the African Plio-Pleistocene. *J. Hum. Evol.*, **32**, 289–322.

Riedman, M.L. & Estes, J.A. (1987). A review of the history, distribution and foraging ecology of sea otters. In *The Community Ecology of Sea Otters*, ed. G.R. VanBlaricom & J.A. Estes, pp. 4–21. Berlin: Springer.

Robel, R.J., Dayton, A.R., Henderson, R.F., Meduna, R.C. & Spaeth, C.W. (1981). Relationships between husbandry methods and sheep losses to canine predators. *J. Wildl. Manage.*, **42**, 362–72.

Roelke-Parker, M.E., Munson, L., Packer, C., Kock, R., Cleaveland, S., Carpenter, M., Obiren, S.J., Popischil, A., Hofmann-Lehmann, R., Lutz, H., Mwamengele, G.L.M., Mgasa, M.N., Machanga, G.A., Summers, B.A. & Appel, M.J.G. (1996). A canine distemper virus epidemic in Serengeti lions (*Panthera leo*). *Nature*, **379**, 441–5.

Rombauer, I.S. & Becker, M.R. (1963). *The Joy of Cooking*. London: Dent & Sons.

Rood, J.P. (1986). Ecology and social evolution in the mongooses. In *Ecological Aspects of Social Evolution*, ed. D.I Rubenstein & R.W. Wrangham, pp. 131–52. Princeton, N.J.: Princeton University Press.

Rootsi, I. (1995). *Man-eater Wolves in 19th Century Estonia*. Report. Tartu: Estonian Naturalists Society.

Sacks, J.J., Sattin, R.W. & Bonzo, S.E. (1989). Dog-bite related fatalities from 1979 through 1988. *J. Am. Med. Assoc.*, **262**, 1489–92.

Sargeant, A.B. & Arnold, P.M. (1984). Predator management for ducks on waterfowl production areas in the northern plains. In *Proceedings, 11th Vertebrate Pest Conference*, ed. D.O. Clark, pp. 161–7. Davis: University of California.

Satish, H. (1996). Killings by wolves. In *2nd International Symposium on the Coexistence of Large Carnivores with man, Saitama, Japan*, poster. Tokyo: Ecosystem Conservation Society.

Saunders, G., Coman, B., Kinnear, J. & Braysher, M. (1995). *Managing Vertebrate Pests: Foxes*. Canberra: Australian Government Publishing Service, Bureau of Resource Sciences.

Savage, R.J.G. (1977). Evolution in carnivorous mammals. *Palaeontology*, **20**, 237–71.

Savage, R.J.G. (1978). Carnivora. In *Evolution of African Mammals*. ed. V.J. Maglio & H.B.S. Cooke, pp. 249–67. Cambridge, Mass.: Harvard University Press.

Schaefer, J.M., Andrews, R.D. & Dinsmore, J.J. (1981). An assessment of coyote and dog predation on sheep in southern Iowa. *J. Wildl. Manage.*, **45**, 883–93.

Schaller, G.B. (1967). *The Deer and the Tiger*. Chicago: University of Chicago Press.

Schaller, G.B. (1972). *The Serengeti Lion*. Chicago: University of Chicago Press.

Schaller, G.B. (1996). Carnivores and conservation biology. In *Carnivore Behavior, Ecology and Evolution II*, ed. J.L. Gittleman, pp. 1–10. Ithaca, N.Y.: Cornell University Press.

Schaller, G.B., Hu, J., Pan, W. & Zhu, J. (1985). *The Giant Pandas of Wolong*. Chicago: University of Chicago Press.

Schaller, G.B. & Lowther, G.R. (1969). The relevance of carnivore behavior to the study of early hominids. *Southwest. J. Anthropol.*, **25**, 307–41.

Scrivner, J.H., Howard, W.E., Murphy, A.H. & Hays, J.R. (1986). Sheep losses to predators on a California USA range 1973–1983. *J. Range Manage.*, **38**, 418–21.

Seidensticker, J., Christie, S. & Jackson, P. (1999). *Riding the Tiger.* Cambridge: Cambridge University Press.

Seip, D.R. (1992). Limiting factors of woodland caribou and interrelationships with wolves in southeastern British Columbia. *Can. J. Zool.*, **70**, 1494–503.

Serpell, J.A. (1988). The domestication and history of the cat. In *The Domestic Cat,* ed. D.C. Turner & P. Bateson. pp. 151–8. Cambridge: Cambridge University Press.

Shipman, P. (1986). Scavenging or hunting in early hominids: theoretical framework and tests. *Am. Anthropol.*, **88**, 27–43.

Short, J., Bradshaw, S.D., Giles, J., Prince, R.I.T. & Wilson, G.R. (1992). Reintroduction of macropods (Marsupialia, Macropodoidea) in Australia: a review. *Biol. Conserv.*, **62**, 189–204.

Sidorovich, V., Kruuk, H., Macdonald, D.W. & Maran, T. (1998). Diets of semiaquatic carnivores in northern Belarus, with implications for population changes. *Symp. Zool. Soc. Lond.*, **71**, 177–89.

Siebold, W. (1959). *Die Wildküche.* Giessen: Brühlischer Verlag.

Sillero-Zubiri, C. & Macdonald, D. (1997). *The Ethiopian Wolf – Status Survey and Conservation Action Plan.* Gland: IUCN.

Siminski, P. (1998). Mexican wolf: conservation and reestablishment in the wild. http://www.desertmuseum.org/conservation/mexicanwolf.html

Sinclair, A.R.E. (1979). Dynamics of the Serengeti ecosystem: process and pattern. *In Serengeti, Dynamics of an Ecosystem,* ed. A.R.E. Sinclair & M. Norton-Griffiths, pp. 1–30. Chicago: University of Chicago Press.

Sinclair, A.R.E. (1985). Does interspecific competition or predation shape the African ungulate community? *J. Anim. Ecol.*, **54**, 899–918.

Sinclair, A.R.E. & Pech, R.P. (1996). Density dependence, stochasticity, compensation and predator regulation. *Oikos*, **75**, 161–73.

Sovada, M.A., Sargeant, A.B. & Grier, J.W. (1995). Differential effects of coyotes and red foxes on duck nest success. *J. Wildl. Manage.*, **59**, 1–9.

Stander, P.E. (1992). Cooperative hunting in lions: the role of the individual. *Behav. Ecol. Sociobiol.*, **29**, 445–54.

Stanford, C.B. (1999). *The Hunting Apes.* Princeton, N.J.: Princeton University Press.

Stearman, A.M. & Redford, K.H. (1992). Commercial hunting by subsistence hunters – Siriono Indians and Paraguay Caiman in lowland Bolivia. *Hum. Organ.* **51**, 235–44.

Strachan, R. & Jefferies, D.J. (1993). *The Water Vole Arvicola terrestris in Britain 1989–1990: Its Distribution and Changing Status.* London: Vincent Wildlife Trust.

Swenson, J.E., Sandgren, F., Heim, M., Brunberg, S., Sørensen, O.J., Söderberg, A., Bjärvall, A., Franzén, R., Wikan, S., Wabakker, P. & Overskaug, K. (1996). Is the Scandinavian brown bear dangerous? *NINA Oppdragsmeld.*, **404**, 1–26.

Tapper, S.C., Potts, G.R. & Brockles, M.H. (1996). The effect of experimental reduction in predation pressure on the breeding success and population density of grey partridges *Perdix perdix. J. Appl. Ecol.*, **33**, 965–78.

Thiess, A., Schuster, R., Nockler, K. & Mix, H. (2001). Helminth findings in indigenous racoon dogs *Nyctereutes procyonides* (Gray, 1834). *Berl. Munch. Tierarztl. Wochenschr.*, **114**, 273–6.

Tinbergen, N., Broekhuysen, G.J., Feekes, F., Houghton, J.C.W., Kruuk, H. & Sculz, E. (1965). Egg-shell removal by the black-headed gull, *Larus ridibundus* L., a behavioural component of camouflage. *Behaviour*, **19**, 74–117.

Trout, R.C. & Tittensor, A.M. (1989). Can predators regulate wild rabbit *Oryctolagus cuniculus* population density in England and Wales? *Mamm. Rev.*, **19**, 153–74.

Tsukahara, T. (1993). Lions eat chimpanzees: the first evidence of predation by lions on wild chimpanzees. *Am. J. Primatol.*, **29**, 1–11.

Turner, A. (1985). Extinction in large African carnivores. *S. Afr. J. Sci.*, **81**, 256–7.

Turner, A. (1990). The evolution of the guild of larger terrestrial carnivores during the Plio-Pleistocene in Africa. *Geobios*, **23**, 349–68.

Van Valkenburgh, B. (1999). Major patterns in the history of carnivorous mammals. *Annu. Rev. Earth Planet. Sci.*, **27**, 463–93.

Vander Wal, S.B. (1990). *Food Hoarding in Animals*. Chicago: University of Chicago Press.

Varty, K. (1967). *Reynard the Fox*. Leicester: Leicester University Press.

Veitch, C.R. (1985). Methods of eradicating feral cats from offshore islands in New Zealand. In *Conservation of Island Birds*, ed. P.J. Moors, pp. 125–42. ICBP Tech. Publ. 3. London: International Council for Bird Preservation.

Venkataraman, A.B. (1995). Do dholes (*Cuan alpinus*) live in packs in response to competition with or predation by large cats? *Current Sci.*, **69**, 934–6.

Venkataraman, A.B., Arumugam, R. & Sukumar, R. (1995). The foraging ecology of dhole (*Cuan alpinus*) in Mudumalai Sactuary, southern India. *J. Zool.*, **237**, 543–61.

Vila, C., Savolainen, P., Maldonado, J.E., Amorim, I.R., Rice, J.E., Honeycutt, R.L., Crandall, K.A., Lundeberg, J. & Wayne, R.K. (1997). Multiple and ancient origins of the domestic dog. *Science*, **276**, 1687–9.

Vinnicombe, P. (1976). *People of the Eland*. Pietermaritzburg: University of Natal Press.

Vrba, E.S. (1985). Environment and evolution: alternative causes of the temporal distribution of evolutionary events. *S. Afr. J. Sci.*, **81**, 229–36.

Vrba, E.S. (1988). Late Pliocene events and hominid evolution. In *Evolutionary History of the Robust Australopithecines*. ed. F.E. Grine, pp. 405–26. Cape Town: Aldine de Gruyter.

Vucetich, J.A. & Creel, S. (1999). Ecological interactions, social organization, and extinction risk in African wild dogs. *Conserv. Biol.*, **13**, 1172–82.

Walker, A. (1984). Extinction in hominid evolution. In *Extinctions*, ed. M.H. Nitecki, pp. 54–64. Chicago: University of Chicago Press.

Walther, F.R. (1969). Flight behaviour and avoidance of predators in Thomson's gazelle (*Gazella thomsonii* Guenther 1884). *Behaviour*, **34**, 184–221.

Ward, J.P. & van Dorp, D.A. (1981). The animal musks and a comment of their biogenesis. *Experientia*, **37**, 917–35.

Weber, J.M. & Roberts, L. (1990). A bacterial infection as a cause of abortion in the European otter, *Lutra lutra*. *J. Zool.*, **219**, 688–90.

Werdelin, L. (1996). Carnivoran ecomorphology: a phylogenetic perspective. In *Carnivore Behavior, Ecology and Evolution II*, ed. J.L. Gittleman, pp. 582–624. Ithaca, N.Y.: Cornell University Press.

Werdelin, L. & Solounias, N. (1991). The Hyaenidae: taxonomy, systematics and evolution. *Foss. Strata*, **30**, 1–104.

Western, D. & Henry, W. (1979). Economics and conservation in third world national parks. *BioScience*, **29**, 414–8.

White, P.C.L., Gregory, K.W., Lindley, P.J. & Richards, G. (1997). Economic values of threatened mammals in Britain: a case study of the otter *Lutra lutra* and the water vole *Arvicola terrestris*. *Biol. Conserv.*, **82**, 345–54.

White, T.H. (1976). *The Book of Beasts, a Translation from a Latin Bestiary*. London: Cape.

Winnikof, L. (1995). Questionnaire survey of visitors to the Kruger National Park. Unpublished Report. Skukuza: Kruger National Park.

Woodburn, J. (1968). An introduction to Hadza ecology. In *Man the Hunter*, ed. R.B. Lee & I. De Vore, pp. 49–55. Chicago: Aldine.

Woodroffe, R. & Ginsberg, J.R. (2000). Ranging behaviour and vulnerability to extinction in carnivores. In *Behaviour and Conservation*, ed. L.M. Gosling & W.J. Sutherland, pp. 125–40. Cambridge: Cambridge University Press.

Woodroffe, R., Ginsberg, J. & Macdonald, D. (1997). *The African Wild Dog – Status Survey and Conservation Action Plan*. Gland: IUCN.

Woodroffe, R., Macdonald, D.W. & da Silva, J. (1995). Dispersal and philopatry in the European badger (*Meles meles*). *J. Zool.*, **237**, 227–39.

Wyatt, A.W. (1950). The lion men of Singida. *Tanganyika Notes Rec.*, **28**, 45–53.

Youngman, P.M. (1982). Distribution and systematics of the European mink *Mustela lutreola* Linnaeus 1761. *Acta Zool. Fenn.*, **166**, 1–48.

Zagers, J.J.A. & Boersema, J.H. (1998). Infections with *Baylisascaris procyonis* by humans and raccoons. *Tijdschr. Diergeneeskd.*, **123**, 471–3.

Index